BLACK AND WHITE
IDENTITY FORMATION

BLACK AND WHITE IDENTITY FORMATION

STUART T. HAUSER, M.D., Ph.D.
EYDIE KASENDORF, M.A.

SECOND EDITION

ROBERT E. KRIEGER PUBLISHING COMPANY
MALABAR, FLORIDA

Original Edition 1971
Second Edition 1983

Printed and Published by
ROBERT E. KRIEGER PUBLISHING COMPANY, INC.
KRIEGER DRIVE
MALABAR, FLORIDA 32950

Copyright © (Original Material) 1971 by
John Wiley & Sons, Inc.
Transferred to Stuart T. Hauser
Copyright © (New Material) 1983 by
ROBERT E. KRIEGER PUBLISHING CO., INC.

Printed in the United States of America

Library of Congress Cataloging in Publication Data

Hauser, Stuart T.
  Black and white identity formation.

  Bibliography: p.
  Includes index.
  1. Identity (psychology) 2. Socially handicapped youth—
Psychology. 3. Afro-Americans—Psychology. 4. Caucasian
race—Psychology. 5. Adolescent psychology.
I. Kasendorf, Eydie. II. Title.
BF724.3.13H38 1983   155.55   82-16221
ISBN 0-89874-055-X

To Our Families

PREFACE TO THE FIRST EDITION

This monograph presents, discusses, and offers approaches to understanding the results of longitudinal studies of white and black adolescents. Multiple facets of method, data, and theoretical models are in turn taken up within the following chapters. However, two important issues are not directly addressed within the text, although they are alluded to in many places. These issues touch upon the interpretation and implications of our findings. In these preliminary notes, I raise the problems and thereby encourage the reader to keep them in mind as he then proceeds through the more detailed theory, method, and data.

The issues are related to empirical concerns. First of all, there is the matter of the *nature* of the sample. The group of adolescents that were longitudinally studied was a small one. It was of sufficient size to permit meaningful application of nonparametric statistical tests. The results of these tests, graphical techniques, and clinical approaches certainly argue for meaningful differences *within* the sample. As the same time, it is crucial to remember that these are findings derived from a small matched interracial sample. They are *not* results that in any way, by themselves, typify a "black ghetto" or "white slum." Such a leap in reasoning is ever tempting, and often fallacious. Nonetheless, even given this note of caution we are left with a serious problem: namely, how are we to understand findings that show the black adolescents to have different identity formation patterns from white adolescents. Much of the monograph is devoted to considering this extremely fascinating and perplexing question.

A second issue is the era within which this study was carried out. These adolescents were followed between 1962 and 1967. The many events since then — ranging from major civil rights demonstrations to new organizations and assassinations — may have generated significant impacts upon those aspects of the sociocultural surroundings important for adolescent development. Thus contemporary white and black adolescents might show very different identity patterns from those within our sample. This is an important kind of speculation, for it is directed toward attempting to understand the developmental impacts of these seemingly major societal events.

To complete a longitudinal study of this kind required the support and cooperation of many people, most of all the subjects themselves. Their names have been disguised within the text for the sake of confidentiality. This does not detract from the debt I think the investigation owes to them for their sustained cooperation in allowing us to observe them growing up.

A number of other people made significant contributions to the planning, carrying out, and completion of this investigation. Much of the original planning and testing was done while I was a medical student at Yale University School of Medicine. My advisor in these phases was Dr. Ernst Prelinger. His teaching, guidance, and support was crucial in enabling the study to be conceived and then taken through the first critical stages of data collection and analysis.

In the later stages of the work, the further analysis and understanding of results, Dr. Elliot Mishler was most instrumental. Our many discussions, his careful reading of the manuscript, his many suggestions about analysis and display of data represent but a portion of his contribution. Indeed, his intellectual and personal support of my work is most directly responsible for this monograph.

While working with Dr. Mishler in the social psychiatry laboratory of Massachusetts Mental Health Center, a number of other members of the laboratory were extremely helpful. Dr. Nancy Waxler read early portions of the manuscript and offered valuable suggestions concerning statistical techniques and tables. Members of the Research Training in Social Psychiatry seminar were helpful in discussing and questioning aspects of the study; I am grateful for their stimulating questions and interest to H. Stephen Leff, Kent Ravenscroft, Virginia Abernathy, Eve Bingham, and Ron Walter.

I am, in addition, greatly indebted to Dr. Jack Ewalt, Superintendent of Massachusetts Mental Health Center and Professor of Psychiatry at Harvard Medical School, for his continued encouragement and support of this research over the last two-and-one-half years.

More specifically, I am indebted to Drs. Kenneth Keniston, Florence Shelton, Peter Reich, John Mack, and George Vaillant for several valuable discussions about facets of the study. In addition, the writings of Erik Erikson, David Reisman, and Lee Rainwater have influenced much of the design and interpretation of the investigation. The sensitive and intelligent comments of Dr. Irving B. Weiner, the series editor, were of much assistance in shaping the final form of the monograph.

Finally, I want to acknowledge the help of several people in the many and tedious tasks involved in performing the studies. Charles Twyman, Grace Dowdy, Arnold Lerner, Gerald Barberisi, Nicholas Defeo, Hildegarde Schwartz, and Robert Schreck were most helpful to me in obtaining research subjects and neighborhood facilities in New Haven. Richard Cash and Susan Bender were helpful in early recording work.

Roger Bakeman, with his intelligent help in programming, allowed the data to be analyzed through the Yale Computer Center. Two research assistants, Kathleen McCormack and Gerry Pilgrim, were most helpful through their stimulating questions and practical assistance in the last revisions of the manuscript. A series of secretaries have contributed in flexible and efficient ways to the several versions of the manuscript: Anne Granger, Judy Wettenstein, Jacqueline Hooker and, in the final important stages, Nancy Fuller. I am equally grateful to Linda Twing who typed the stencils for the final copy.

I am indebted to two sources for financial support of the project: in the summers of 1962-1965 Yale University School of Medicine provided National Institute of Health summer research fellowships. Foundations Fund for Research in Psychiatry provided a small grant to aid in the final collection and transcription of interview data.

Last, but certainly not least, I am greatly indebted to my wife, Barbara Hauser. In all phases of this project, she has helped directly with her intelligent discussions and editing. Even more important has been the less tangible aspect of her assistance; through her patience and support of the time and energy the project demanded of her husband. Her role in these many facets of the work is most appreciated and really reflected throughout the monograph.

Stuart T. Hauser

Boston, Massachusetts
June, 1970

PREFACE TO THE SECOND EDITION

Over a decade has passed since the completion of the study which was described and discussed in *Black and White Identity Formation*. Within these years, new understandings, analyses, and observations about black personality development have been reported. The already large literature in this area continues to expand in many directions.

In the original preface, and then at greater length in the final chapters, several implications and interpretations of the findings were taken up. The recent empirical studies and theoretical analyses now raise many questions about the explanatory models which we applied to the original data. Alternative approaches for understanding black families and personality development are prominent within conceptual as well as empirical studies.

The objective of this second edition is to examine results of this study once again, in light of the new contributions in this area from psychologists and other social scientists. All of the original chapters have been rewritten, some extensively, in order to incorporate new conceptualizations and data. In working on the second edition we realized that a change in the sequence of chapters would also contribute to the clarity of the monograph. Finally, two new chapters have been added which review the extensive empirical and conceptual literature which has appeared since 1970.

It is in these chapters that we explicitly reconsider some of the initial interpretations formulated in the first edition of *Black and White Identity Formation*.

The conceptual and empirical shifts in the field have obviously influenced the thinking of the senior author. His involvement in a critical review of racism studies (Hauser, 1973), followed by a dissertation studying several aspects of interracial interviews (1976), brought considerable new breadth to Hauser's understandings of black personality and development. A greater immersion in studies of adolescent ego development and family interaction (Hauser, 1980; Hauser, *et al.*,

1979) as well as completion of graduate training in psychoanalysis and psychology brought still other fresh perspectives and sensitivities to his work. Yet another key influence in the evolution of this new edition has been the addition of a new collaborator, the second author. Through her energetic and thoughtful work, much of the literature review and critique was completed. Ms. Kasendorf was also most helpful in her many observations about the original chapters as we rethought them for the second edition. Presentation and discussion of some of these reconsiderations at a symposium on "Prejudice and Psychopathology" at Cornell University Medical College in May, 1979 helped to crystallize many of the ideas and approaches in this new edition.

In the final stages of preparing this edition, Joyce Olson offered unsolicited literature and editorial suggestions which were extremely helpful. We are grateful for this unanticipated contribution to our task. All of the final typing was carried out diligently and competently by Margot Hubbard.

The open and stimulating atmosphere of the Laboratory of Social Psychiatry, and many discussions over the years with Elliot Mishler, provided significant but not easily specified support for this undertaking. More specific financial support was largely through an NIMH Research Scientist Development Award (#5K01MH70178) to Dr. Hauser. Robert Krieger, of Krieger Publishing, originally brought up the idea of this second edition and then patiently waited as we completed this unexpectedly complex, and unusually interesting task. We are grateful to him for his initial suggestion and continued support along the way.

Finally, we are indebted to our families for their patience in enduring the many hours of writing and talk which was required for us to complete this large project. Ms. Kasendorf thanks John Quinlivan for his ongoing encouragement in this first major publication of her professional career. For this new edition, Ethan and Joshua Hauser were now considerably older and more aware of the demands which such a project placed upon their father and family. Barbara Hauser once again was supportive and encouraging in this new phase of an old project.

Stuart T. Hauser
Eydie Kasendorf

Boston, Massachusetts
April, 1980

Contents

CHAPTER 1

An Overview

INTRODUCTION

The study presented was conducted between 1962 and 1967. The theoretical context of the study was shaped by the plethora of writings that addressed themselves to describing and analyzing the lives of black Americans. We noted then that articles and studies of all conceivable kinds were flooding bookshelf and newsstand, ranging from surveys of thousands of blacks and whites to detailed investigations of specific historic, economic, and political ramifications of black American problems.[1] In the decade since publication of our original findings, this literature has grown considerably and taken on new emphases. The shifts in the writings are towards new conceptual frameworks and empirical directions. In the first chapter we review the theoretical perspectives which were first reflected in our approach to studying black and white identity formation. In addition, we take up new theoretical and empirical writings about identity formation. Following elaboration of our own study, in subsequent chapters we return to summarize and discuss new personality studies and theoretical formulations which address the adaptation of black Americans.

THEORY

When this study was first conceived and conducted, rigorous psychological studies of black Americans were considerably outnumbered by investigations which dealt with historical and sociopolitical dimensions of race relations. Observations and tests of black children in the midst of school desegration, and social psychological perspectives on black Americans were among the more outstanding contribu-

---

[1]See, for example, "The Negro in America," *Newsweek*, July 1963, and *Look*, January 7, 1969, for such surveys. More detailed studies are "The Negro American," *Daedalus* (1965), and "The Negro Protest," *Annals of the American Academy of Political and Social Sciences* (1965).

1

tions in this genre.[2] Our investigation also had roots in the difficult and perplexing current issues of American race relations. Yet, it differed in an important way from most of the other studies. While begun before Pettigrew's plea for "theoretically guided" research on black problems, it was with this orientation in mind that this study was designed and undertaken.[3]

The theoretical interest in forming the study is in the relationship between personality development and sociocultural context.[4] Careful and detailed studies of social and cultural systems have been present within the social science literature for many years. Similarly, although with different methods, psychoanalytic investigators have attempted systematic study of individual development, of personality systems, and of personality structures. A most difficult task has remained: the conceptualization and empirical study of the relationships between these understandings of psychological development and of social reality.

Early approaches to this interface problem were taken by Heinz Hartmann and his associates.[5] These works were then followed by those of Erik Erikson, who concentrated more specifically on the various phases of development, ranging from infancy to old age. Erikson's theory outlines a sequence of phases of psychosocial development and relates these phases to psychosexual development. In each phase of the individual life cycle there is a phase-specific developmental task with derivative conflicts, which must be solved by the growing individual for his or her psychological maturation. Funda-

[2]See R. Coles, *Children of Crisis* (1967); T. Pettigrew, *Profile of the American Negro* (1964); and, *Black Rage* (1968)—the more clinical and social criticism by W.H. Grier and P.M. Cobb.

[3]T.F. Pettigrew, "Negro American Personality: Why Isn't More Known?," *Journal of Social Issues*, 20, 1964.

[4]Representing interest in 1970 were writings by B. Kaplan (Ed.), *Studying Personality Cross-Culturally* (1961); C. Dubois, *The People of Alor* (1944); E. Erikson, *Childhood and Society* (1950); H. Hartmann, E. Kris, and P.M. Lowenstein, "Some Psychoanalytic Comments on Culture and Personality," in *Psychoanalysis and Culture*, G. Wilbur (Ed.) (1949). The above all represent the initial and continuing interest in such problems. Additional contributions include Erikson's *Young Man Luther* (1958); *Identity: Youth and Crisis* (1968); and *Ghandi's Truth* (1969); K. Keniston's *The Uncommitted* (1966); S.M. Elkin's *Slavery* (1963); R.J. Lifton's *Thought Reform and the Psychology of Totalism* (1961); and the anthology edited by R.W. White, *The Study of Lives* (1963).

[5]H. Hartmann, *Ego Psychology and the Problem of Adaptation* (1939) and Hartmann, Kris, and Lowenstein, *op. cit.*, are two of the best known statements of this early approach.

mental to this theory of development is its

> . . . conceptual explanation of the individual's social develop-
> ment by tracing the unfolding of the *genetically social char-*
> *acter* of the human individual in the course of his encounters
> with the social environment at each phase of the epigenesis.[6]

Of further importance is the theory's position that " . . . the society into
which the individual is born makes him a member by influencing the
manner in which he solves the tasks posed by each phase of his
epigenetic development."[7] The phases of development are conceived
of as universal. The typical solutions to the developmental tasks are
conceived of as varying from society to society.

Adolescence, long a major area of interest for students of human
development, is now a prominent focus for social scientists as well.
The literature of adolescence reflects the dual perspectives embodied
in these two approaches:

> One [approach] lays its stress on the more individual aspect of
> the problem, as in psychoanalysis; the other regards the
> individual adolescent as a member of society, and considers
> his actions, his problems, even his aspirations to be the resul-
> tant of forces which have their origin in social developments.[8]

With the theoretical formulations by Hartmann and Erikson, the
potential dichotomy between these two approaches has lessened.
Indeed, it is now clear that there are highly significant sociocultural
aspects to individual adolescent development. The grounds for this
argument are both theoretical and empirical.[9]

Continuing work in the direction of further careful understanding
of the interface between sociocultural context and adolescent development is the particular orientation of this study. Our theoretical
position is closely tied to Erikson's view of "psychosocial develop-

---

[6]D. Rapaport, "A Historical Survey of Psychoanalytic Ego Psychology," *Psychological
Issues*, 1, 15, 1959. Italics are Rapaport's.

[7]*Ibid.*

[8]Deutsch, H., *Selected Problems of Adolescence* (1967), p. 10.

[9]To some extent all of the works cited in footnote 4 present evidence for this position.
It is discussed further by Deutsch, H., *op. cit.* In addition, the anthologies by Brody, E.,
Minority Group Adolescents in the U.S. and Dragstin and Elder (1975) contain pertinent
papers.

ment." In his theoretical model, the central task of adolescence is that of *identity formation*. In this perspective, concepts at the core of and radiating from this task are thus basic to the investigation of adolescence.

During adolescence, problems of coherence, continuity, meaningfulness, and self-definition may and frequently do take precedence in individual awareness. At times these problems take on overwhelming importance. It is within this period of development that correspondences between childhood expectations and envisioned adulthood are then sought, and frequently not found.

The years preceding adolescence are ones of relative calm. Oedipal conflicts have for the moment subsided. Sexual impulses are, again for the moment, dormant. This is a time of much practical learning. The school age child is learning skills which are directly introducing him to the complex technology and thought of his surrounding society. Parents and teachers, both formal and informal are seen increasingly as being among the crucial representatives of society. In other words, emergent at this point is a growing awareness of roles and relationships outside of the immediate family setting.

With the advent of puberty there occur major shifts in psychological as well as physical development. Previously established ideals, rules, defenses, perceived continuities are now rearranged, or rejected.[10] Besides the internal biological stresses of intensified sexual drives and rapidly advancing bodily changes, numerous pressures of a social nature become crucial. These include issues of new physical and psychological intimacies. And present as well are those issues related to the social setting itself: the choices of occupational role, social standing within the local community, and the heightened presence of the peer group. The young person experiences a seeming urgent pressure to make what appear to be irreversible commitments: commitments which are personal, sexual, occupational, and at times ideological. Within this turmoil self-conscious and disquieting questions of adolescence become more conscious: "What do they think of me? Do they like me? What do I stand for? Where do I belong?" Definition, synthesis, continuity, are the leading themes of this turbulent time:

> What the regressing and growing, rebelling and maturing youth are now primarily concerned with is *who* and *what* they are in the eyes of the wider circle of significant people as compared to what they themselves have come to feel they

---

[10]This general problem of adolescent shifts and "rearrangement" is discussed by Freud, A., *The Ego and Mechanisms of Defense* (1948).

are; and how to connect the dreams, idiosyncrasies, roles, and skills cultivated earlier with the occupational and social prototypes of the day.[11]

...the ego of the adolescent is in great need of support, yet paradoxically it has to provide this support out of its own resources. Against newly intensified impulses it has to maintain the old defenses and create new ones; it has to consolidate achievements that have already been reached. The most important of its tasks is the struggle to synthesize all childhood identifications as they become enlarged and enriched by new ones. The successful end result of this struggle will be the formation of a solidified personality, endowed with a subjective feeling of identity that is confirmed and accepted as such by society.[12]

These, then, are the important and specific kinds of adolescent development problems, generated by an intertwining of intrapsychic and social forces. They are the problems of identity formation, the developmental issues to which this study is addressed.

The conceptual framework underlying this study is that of psychoanalytic ego psychology. The overall perspectives and questions posed are to a large extent guided by psychoanalytic formulations about adolescence. More specifically, the choice of hypotheses and problems is based on the discussions of Erikson (1956, 1958, 1968). In clinical and theoretical studies the notion of identity formation is basic to many of his analyses of psychosocial issues, most particularly those of adolescence. Identity formation represents the organization of a number of intrapsychic and psychosocial components, a synthesis encompassing a wide array of ego functions:

...an evolving configuration of constitutional givens, idiosyncratic libidinal needs, favored capacities, significant identifications, effective defenses, successful sublimations, and consistent roles.[13]

It is the formation of this configuration that is seen by Erikson as both a focus for and source of critical psychosocial problems in adolescence.

On several occasions Erikson (1950, 1958, 1959, 1968) discusses the theoretical implications and empirical referents of identity formation.

[11]Erikson, E., *Growth and Crises of the Healthy Personality* (1960).
[12]Deutsch, H., *op. cit.* p. 33.
[13]Erikson, E., "The Problem of Ego Identity," *Psychological Issues*, 1956, p. 116.

Repeatedly, he and others working in this area have noted theoretical issues such as the relationship of identity to the older more systematized psychoanalytical concepts of ego, ego ideal, super-ego, identification, and self. In a recent review Abend (1971) carefully discusses some of these questions and presents several helpful solutions. There is also the important problem of how identity formation is linked with non-psychoanalytic concepts. What, for example, are the distinctions between ego identity, identity formation and role, self-image, persona, and lifestyle? This is an especially significant problem, for some of the issues referred to by identity formation and ego identity overlap with those of the social psychologist, for whom role, self systems, and other such terms are central. Clearly, if there is not to be duplication of effort and loss of needed comparative data, methods of "translation" or reconciliation of these concepts is needed. The problem is particularly apparent, since major research about black personality development, is focussed on self-esteem and self-concept.

In addition to these conceptual issues, there are a host of empirical questions raised by the notion of identity formation. In general, systematic empirical research about identity development has drawn from a very special minority group; white, middle socioeconomic class American youth (usualy college students).[14] Such a sampling bias has restricted legitimate generalizations about identity processes as well as empirical testing of underlying assumptions. There are multiple questions which must be asked about identity formation in other social and cultural groups. Is the process a significant one in individuals of all populations? If it is, do identity development problems nonetheless take different forms and find very different solutions in other groups? Is identity development the predominant task of adolescence in all sociocultural groups? Or does identity formation emerge as a critical problem in other development periods for individuals of differing sociocultural contexts? None of these questions can satisfactorily be

---

[14]See, for instance, Blaine and McArthur, *Emotional Problems of the Student* (1961); Wedge, B., *Psychological Problems of College Men* (1958); Blos, P., *On Adolescence* (1962). Prelinger (1962) has suggested several reasons for this preferred sample: 1) identity development occurs over a relatively longer period of time for college students; 2) college students are confronted with both a wider range of possible career choices as well as social and ideological commitments; 3) college students usually have a greater variety of intellectual, social, and cultural resources available to them allowing "the development of more differentiated and various ego identities"; 4) college students are easily available for study; and 5) they speak a language understood by identity investigators, who have had a relatively similar background to the college students.

Studies by Erikson (1950, 1958), Lifton (1962), Ross (1962), Prelinger (1958), and Matteson (1974) represent exceptions to this bias.

answered until identity studies draw their sample from a greater number and variety of human populations. And neither can the two fundamental assumptions implied by the concept of identity formation and embodied in the preceding questions be validated, namely:

1. that formation of identity is a developmental problem found in all social and cultural groups, and though the process may be exacerbated or foreshortened, its occurrence is nevertheless not "culture-bound,"
2. that formation of identity emerges as a critical psychosocial problem in late adolescence. (Studies in the United States suggesting that this period is in the late teens or early twenties.)[15]

Specific formulations and their implied predictions about identity formation will be detailed later. This survey of the issues raised by the concept suggests the orientation taken by our study. It is a study most avowedly intended as an exploration of identity development in varied sociocultural contexts. With this end in mind, working class black and white boys were chosen as research subjects, to be followed closely over a three-year period. Guiding the research was Erikson's description of identity formation outlined above in its most complicated and ambiguous form. Though possibly of clinical usefulness in this form, the description requires several changes in the direction of greater explicitness and empirical specification for non-clinical investigative purposes. The result of these changes is a more "operational" definition, which will be discussed and presented in subsequent chapters.

Dignan proposes a definition of identity. She suggests ego identity is, "the complex of self-referent images which evolves through social interaction, thereby delimiting the self."[16] Although commendable for its relative clarity, her definition lacks emphasis on certain processes which are crucial to the notion of identity formation, namely *continuity* and *synthesis*. Moreover, the unconscious aspects of identity are ignored by her definition. Sister Dignan is one of several students who have addressed themselves to the empirical study of identity formation. In addition to the obvious relevance of these empirical identity studies,

[15]It is of course difficult to define adolescence and late adolescence cross-culturally. The start of the period might be viewed biologically as the onset of puberty, as the time of certain sexual and general bodily changes. But what late adolescence means remains a problem. One possible solution might be to avoid the term "late adolescence," and speak instead of amount of time since puberty. This would emphasize the possibility that this period, and perhaps the other developmental ones as well, has a large degree of variation in extent and form within the limits imposed by certain biological constants.

[16]Dignan, M.H., "Ego Identity and Maternal Identifications," *Journal of Personality and Social Psychology*, 1:476, 1965.

also pertinent to the research reported here are those works dealing with social class, the American black, and selected aspects of adolescent development. Besides sources from the social sciences and psychoanalysis, there are a number of relevant literary contributions. Following a selective review of this literature,[17] we will return to the current study, then taking up in greater detail the basic concepts and methodological underpinnings.

RELATED PERSPECTIVES

Several of Erikson's original descriptions and clinical discussions of ego identity appear in *Childhood and Society* (1950), *Psychological Issues* (1959), and *Identity: Youth and Crisis* (1968). The latter collection includes papers which are rich in theoretical and clinical considerations, displaying as it were the immense complexity of both the identity construct and the events — psychosocial and intrapsychic — which he is attempting to isolate.[18] The meaning of "ego identity" and "identity formation" is thus greatly expanded by example and usage, resulting in much intuitive understanding of the concept. Much of the difficult work of further conceptual and methodological analysis has been taken up by other students, surely stimulated by the wealth of clinical materials offered by Erikson and his associates.[19]

Predominantly theoretical identity studies, with clinical illustration, include those of Blos (1962), Greenacre (1958), Jacobson (1964), Abend (1971), Wheelis (1958), Lynd (1960), and Strauss (1959). The first four authors discuss implications of ego identity in terms of its place within psychoanalytic theory and often vigorously attack the concept for its many ambiguities. This is particularly so for Jacobson, who in *The Self and the Object World*, points to what she considers a confusion of subjective and objective identity in Erikson's discussions. The concept of the "self" continues to be the major focus of her monograph, identity being treated as it relates to this elusive, perhaps even more complex and ambiguous, concept. Greenacre, Blos, and Abend, on

---

[17]Such a review here must be limited and will therefore concentrate on those works most pertinent to the issues being pursued here. For instance, there have been many recent discussions of identity formation some of which were mentioned above. Those of particular theoretical or methodological importance are surveyed here. A more complete listing of identity studies will be found in the bibliography, and in the recent reviews by Marcia (1980) and Bourne (1978a, 1978b).

[18]Some of the many insights and propositions of these writings have already been noted. More extended discussions of these, and other assumptions and implied hypotheses are taken up in subsequent chapters.

[19]See Erikson (1962), Keniston (1965), Lifton (1961), Prelinger (1958), Blaine (1961), Strauss (1963), Wedge (1958), Wheelis (1958), and White (1964).

the other hand, are directly interested in the concept of identity. Greenacre is primarily concerned with the subjective "sense of identity," stressing the dual aspects of uniqueness, differentiation from all others, and continuity:

> [identity means] . . . an individual or object whose component parts are sufficiently well integrated in the organization of the whole that the effect is a general oneness, a unit.[20]

Less psychoanalytically oriented is Helen Lynd's book, which very sensitively covers many cultural and societal facets of identity formation, drawing upon a broad scope of literary and social science sources. A major portion of the work is devoted to analysis of the relationships of identity formation to issues of shame and guilt. Her notion of identity is wholly derived from Erikson's descriptions, and then enriched by her subtle discussions. To be sure, this does not represent a theoretical or method-oriented study. It is a perceptive and scholarly exploration of cultural and social aspects of identity.

Strauss's study is an example of a social psychologist's approach to identity. From a perspective of social interaction — examining ways in which personal identity, defined primarily in terms of historical continuity, is supported and demolished — Strauss presents lucid theoretical discussions sprinkled with many illuminating clinical illustrations. His emphasis on continuity as a key variable is particularly meaningful. Although the idea is implied in all of his discussions of adolescence, continuity as a central facet of identity is not as clearly elaborated by Erikson. Similar to Strauss's presentation, but much less systematic, is Wheelis's (1958) study of identity. In addition to observations of how sociological and economic changes influence identity development, Wheelis gives detailed and fascinating autobiographical segments as documentation of his theoretical sections. It is through these segments that "clinical definition" of identity is continued.

In addition to the clinical approaches, empirical research in identity formation continues to be of interest to a number of investigators. Three interesting uses of paper-and-pencil tests for identity measurement are given by Scott and Keniston (1959) and Howard (1960), from Erikson's descriptions. Subsequent selections of the test items were made through pre-testing and/or panels of clinicians. Covered by these tests were areas such as time perspective, ideology, role experimentation and sexual identity. Individuals answering the questionnaires were

---

[20]Greenacre, P., "Early Physical Determinants in the Development of the Sense of Identity," *Journal of the American Psychoanalytic Association*, 6, 612, 1958.

from college groups (Keniston and Dignan) and lower middle class teenage girls (Howard). Clinical correlations were attempted in all of the studies. In the Keniston and Scott study, multiple correlations with other paper-and-pencil tests as well as projective tests were determined for each subject. The results, while of uncertain significance, all suggest conformity to general theoretical identity considerations. Thus, for example, Howard found "low identity scores" for "disturbed" girls. Keniston reports high positive correlations between identity scores, other identity scales, and clinical ratings of identity. In addition, Keniston notes negative correlations between identity scores and alienation measures.

In the Keniston study, which was part of a larger project, some of the subjects were followed in subsequent years. Subjects in the other studies were not seen again after the single test administration. Hence, in addition to the caveat regarding the usefulness of this kind of instrument for identity measurement,[21] there is the further problem of whether processes such as those of identity formation can be measured through tests administered at a single sitting. Can an "evolving configuration" be accurately detected through one or more determinations taken at the same point of time? More metaphorically, can a single still picture rather than a movie of someone in motion still accurately capture all the essential parameters of the movement? Our position is that the longitudinal approach *is* required for the study of identity formation.

Marcia (1966, 1967, 1976, 1980) has investigated aspects of identity formation using a structural interview technique which is then coded. Through applying these instruments to samples of undergraduates, he describes four "identity statuses", "styles of resolution of the identity issue" (Marcia, 1980). Several features of his studies differentiate them from the other empirical works. To begin with, Marcia uses the structured interview *and* psychological tests. Moreover, the investigations focus most specifically on different *types* of identity development. And finally, many of the studies undertaken by Marcia and his students use experimental manipulations (e.g., of "self-esteem") as means of differentiating and describing the various "identity statuses." Despite their lack of a longitudinal design, these studies provide a most valuable addition to an obviously limited literature.

They suggest multiple psychological functions which are apparently related to different types of identity formation. Thus, Marcia stimulates attention to the careful empirical determination of identity formation

---

[21]That is, that they reveal the most conscious feelings and attitudes of the subjects, hence avoiding important ego identity dimensions, namely, the unconscious aspects.

and raises questions as to underlying dimensions that might differentiate various forms of identity development. Dimensions of potentially much interest along these lines include cognitive styles, interpersonal patterns, and intrapsychic issues such as self-esteem. The current state of these studies is reviewed in Marcia's recent paper (1980).

There are other empirical studies pertinent to identity formation. Ross (1962) used specially designed TAT pictures to diagnose and analyze identity conflicts among Indonesians in an acculturation situation. Working jointly with an anthropologist, he compared his partially "blind" analyses of the projective test with life history and other field data gathered by the latter. Favorably high correlations were obtained between the test analyses and field data vis-a-vis identity conflicts. The monograph is of particular interest here in that cultural data is used to determine various dimensions and etiologies of the identity issues. A second investigation (Prelinger and Zimet, 1963) employed scales derived from ego psychology considerations. The scales, which include intrapsychic and psychosocial variables, are applied to interview and test materials of college students seen at yearly intervals throughout their college careers.[22] Although again it raises the problem of the sample bias, this approach is most congenial with the one being followed here. Indeed, it meets the "requirements" for identity formation research implied in this short review and critique of the current studies. Alternative measures of identity are reported and discussed by both Marcia (1980) and Bourne (1976a). A thoughtful critical review of the empirical studies is presented by Bourne.

In this section we review the social class research which was most relevant to our original study. The relevant work here can be divided into three general types on the basis of whether a predominantly sociological-anthropological, purely social structural, or psychological perspective is taken.[23] An outstanding example of this first type is Whyte's Street Corner Society (1955). The studies of Walter Miller (1958) have a similar orientation. As in Whyte's work, description of lower class culture is generalized through the observations of small groups in natural settings. But now there are more and larger groups, and the number of observers has increased as well. The position that there exist insulated "sub-cultures" within the United States is implied by Whyte and unequivocally adopted by Miller. A similar assumption of strong coherence and imperviousness to outside influence appears in the

---

[22]Partial description of the study, and extensive rationale for and description of these scales can be found in Prelinger and Zimet (1963). A brief summary of the study is in Prelinger et al. (1960).

[23]In contrast to the preceding section, no attempt has been made to survey the recent large body of studies in these areas.

studies of "teen culture," age-specific patterns of culture said to cut largely across class lines.[24] The existence of such self-sustaining patterns among class or age groups is by no means accepted by all authors. It is especially controversial with regard to "adolescent cultures."[25] At issue is the degree to which these various "sub-cultures" are influenced by other such groupings, and by any values and assumptions possibly held by all groups within the society. There is little question by now of differing cultural patterns within a larger society, patterns which are in part related to socioeconomic distinction.[26]

An anthropological contribution to studies of this type is Elliot Liebow's *Talley's Corner*. For the better part of a year Liebow spent most days, and many nights, with a small group of financially impoverished black men. The lives of this street-corner group are delineated in *Tally's Corner*. Rather than concentrate on "case" or biographical studies, the book focuses on those themes and issues common to "Tally" and his friends. Issues such as "being a man" and "being a father without children" are clarified and explored in much rich detail. Interwoven with these issues is a continuing emphasis on relationships within the group and external to it: The men with one another; the men with their "women"; the men with Liebow, a white Jew; and finally, the men with the surrounding social institutions, ranging from the "carry-out" shop which they frequent, to the employment agencies where they begin interacting with "bosses." Repeatedly, as Liebow so carefully portrays the culture of this group, themes related to both race and social class emerge. Other anthropological approaches here are taken by Abrahams (1964), Hannerz (1969), and Rainwater (1970).

There are a number of more psychological studies concerning lower class youth.[27] Epstein, for example, reports differences in memories between lower and middle class adolescents. The lower class subjects' memories were characterized by increased sexual, aggressive and angry incidents plus a later age of recall (eight years old as compared with three for middle class). Consideration is given in the paper to the influence of the middle class interviewer on both the content and stated age of recall. Korbin also takes up socioeconomic class variables as he reviews the differences in adolescent development between middle and lower class youth. Three "common problems of adolescent development" are considered in both groups of subjects:

---

[24]Bernard (1961), Coleman (1961), Havighurst and Taba (1949).
[25]Elkin and Westley (1955).
[26]See the discussions and studies of Stein and Cloward (1958), Amis (1958), Barber (1961), Fried and Lindeman (1961), Whyte (1955), Hollingshead and Redlich (1958), Myers and Roberts (1959).
[27]Epstein (1963), Korbin (1961), Baittle (1961).

The first of these problems, "adult authority," is resolved temporarily in lower class youth by the absence of adults; thus leading to older peer or sibling supervision, if there is any at all. The solution is only transient, Korbin predicts, for the authority problem readily returns — in unspecified forms — in late adolescence and adulthood. The second problem is the dependence-independence ambivalence. With the shortage of nonparental adults, lower class youth thus have few adults for models. Consequently, they reflect the early independence of the neighborhood, retaining major unfulfilled dependency needs through-out life. Moreover, the absence of accessible adult models leads to the "ideology of the streets" being transmitted relatively intact, rather than that of the family and its specific cultural orientations. Finally, there is the phase-relationship of dependency and autonomy. Lower class youth become autonomous sooner than their middle class counter-parts, in that there is less invested in the former as mobility or "security" objects. But whereas independence strivings are thus accepted more easily, there again is the problem of managing unfulfilled dependency wishes in the lower class adult. The outcome of these resolutions for lower class men is a "primacy of the peer group." Their apparent adult autonomy is marred by a quickened attention given to random types of group support, a tendency which is a direct result of their "unsatisfied psychological hunger for dependency relations."

Because of its clear relevance to this study, Korbin's thoughtful paper has been summarized in some detail. At the time our study was carried out, his review was unique in that it was among the few surveys or studies found which specifically examined relationships between socio-economic class and adolescent psychological development.[28]

The more general area of socioeconomic class and psychopathology or "mental health," has been the object of an increasing amount of

[28]But one can find many papers and books about one special group of lower class youth, .the delinquents. There are abundant studies of their psychopathology, psychotherapy, immediate family context, and social dynamics. Yet, even here, when compared with the attention given to the psychology and sociology of delinquency, there is far less analysis of the relationships between individual delinquent development and sociocultural setting. A striking exception to this is I. Chein's *Road to H* (1964), where relationships between adolescent deviant development and sociocultural matrix are studied in detail. Another exception is the report of the 1955 White House Conference on Delinquency (Witmer 1956), where Erikson, Merton, Redl, and several other sociologists, psychologists, and psychiatrists discuss current knowledge and future research into juvenile delinquency with the concepts and techniques of multiple disciplines. There is also a vast literature in "deviance" where both sociology and anthropology meet in analyzing delinquents, among other "deviants."

popular and scientific interest.[29] It is unclear as to how these primarily epidemiological studies will clarify the relationships and variables which account for correlations between social class and personality development. Using the model of medical epidemiology, we should expect the above works to be most significant in suggesting specific hypotheses about mechanisms underlying the statistical trends which are discovered. Further insight then would be dependent upon the quality of the ensuing clinical and/or experimental studies.

Works on black Americans, our final subject of review, are voluminous. Even when we impose the restriction of direct relevance to questions of personality and culture, their number is still overwhelming. We discuss here those contributions which *preceded* our study. As much as possible the criteria of selection for discussion are quality and overall relevance. Selected relevant works which appeared *after* our study are reviewed in Chapters 8 and 9.

Although there were earlier books,[30] Myrdal's *American Dilemma* (1944) stands as a major early work on the black American and represents the point of departure for most contemporary studies. Dealing largely with historical and institutional facts about blacks, the volumes nonetheless do touch on black class structure and on some personality issues. There is recognition of the tensions inherent in matriarchal family patterns, the degraded self-images, and undischarged aggressions.[31] These subjects are presented by way of anecdotal evidence and general observation, rather than as conclusions of systematic research. Besides psychological comments, Myrdal offers impressively sophisticated insights into the social psychology of American black-white relations:

> The 'American Dilemma' . . . is the ever-raging conflict between on the one hand, the valuations preserved on the general plane . . . the 'American Creed,' where the American thinks, talks, and acts under the influence of high national and Christian precepts, and on the other hand, the valuations on specific planes of individual group living, where personal and local interests; economic social and sexual jealousies; considerations of community prestige and conformity; group prejudice against particular persons or types of people; and all sorts of miscellaneous wants, impulses, and habits dominate his outlook.[32]

[29]Leighton (1959), Srole (1962), Hollingshead and Redlich (1958), Myers and Roberts (1959), Fried and Lindeman (1961), Barber (1961), Kohn (1963).

[30]e.g. Phillips (1981), Dubois (1903).

[31]Several of these "observations," or interpretations are severely challenged in recent discussions, as noted in detail in Chapter 9.

[32]Myrdal, G., An American Dilemma, I:xlvii, 1946.

Since Myrdal, there have been several important historical and sociological studies of black Americans.[33] Elkins' (1963) is of much interest because of his creative use of psychoanalytic and social psychology constructs to reinterpret the system of American black slavery. He sees this system as supporting and/or creating a unique American character type, "Sambo": the obedient, pliant, child-like black. Especially stimulating is Elkins' choice of the Nazi concentration camp as a model for understanding the closed society of American slavery, vis-a-vis total human control. Raised, but left unanswered, is the question of whether this character type has persisted. Has it remained only as a caricature? Or are there still blacks corresponding to the "Sambo" type, either in pure form or as variants from it?[34]

Clark (1965), and Hearn (1962) on a smaller scale, discuss the black ghetto. Drawing upon his work in preparing the Harlem Youth Opportunities (HARYOU) report and the black social psychology literature, Clark presents successive views of sociological, psychological, historic, economic, and political aspects of black ghettoes, in particular, Harlem. Parsons (1965) considers ghetto problems in light of a comparison with the assimilation and inclusion of other minority groups in the United States. His analysis is suggestive. It differs from other discussion of this problem in its argument for the significance of religion as a critical factor distinguishing the black from all other minorities: The black's historically predominant fundamentalist orientation, his total dedication to other worldly concerns, was of great significance in implying to society his "incapacity for full participation." Color was but a symbol for the projection of the social anxiety thereby produced by such an orientation. Although this analysis appears incomplete, and at best controversial, it serves as a strong reminder of the unsolved historical problem: Why have blacks been treated as a minority group in the very specific and disquieting ways in our history? This is also essentially the problem that Elkins pursues. Clearly, social science research into contemporary populations is not going to produce the solution to this great riddle. But, as both Elkins and Parsons demonstrate, the astute application of contemporary psychological and sociological insights may be of unanticipated aid in the reinterpretation of these complex historical processes.[35]

To the historical and sociological dimensions of the American

[33]See Drake and Clayton (1945), Davie (1949), Frazier (1939), Elkins (1959), St. Clari (1965), Parsons (1965), Moynihan (1965), and Clark (1965).

[34]For discussions of this, see Elkins (1963), and Rainwater (1966).

[35]Erikson in his paper, "The Concept of Identity," Race Relations (1965), discussed aspects of American Negro experience and its changing dimensions. The article is a series of observations as opposed to any attempt to discuss black identity formation systematically.

black, Isaacs (1962) adds the parameters of international implications. This is one of the few serious works which confront the question of black identity.[36] Although it is "group identity" which most interests Isaacs, he is repeatedly analyzing issues of individual *and* group black self-images and their vicissitudes under the impact of African and American social changes. The "elements" of black identity are defined to be "name, color, nationality, and origin." Each of the elements are explored in long thoughtful interviews with black literary and political figures and black participants in "Crossroads Africa."[37] Through sensitive and informed analyses Isaacs displays the reciprocal influences present between American blacks and the rest of the world. Toward the end of his discussion he raises the intriguing problem of future development: What will happen as blacks "shed the burdens of nobodiness" and "take on the demands and burdens of somebodiness?"[38] What will be the relationship between the form of future American black group identity and United States society?

Psychological and social psychological studies of American blacks are reviewed in several publications by Pettigrew (1963a, 1964b, 1964c, 1965), and Dreger and Miller (1960, 1968), and Miller and Dreger (1973). There is a broad landscape covered by the many papers in these disciplines, as they range from doll play to IQ-score comparisons, to studies of "self-assimilators". Indeed, one of the serious problems in this area is the very scattering of research efforts frequently leading to either duplication of work or, more often, unsystematic and nontheoretically grounded research.[39] In several discussions, Pettigrew does propose outlines of the theoretical framework which he thinks will be required to account adequately for the social and individual psychology of black Americans. He couches his proposed framework in the language of role theory, which on the whole does not stress historical explanation. The papers emphasize the enormous variation in black behavior. He suggests generic variables and situational variables which, taken together, are said to

[36]Obviously, many of the recent analyses of white "racism" and prejudice address this problem. Three examples of recent work in this area are Jones, *Prejudice and Racism* (1972), Kovel's *White Racism* (1970), and Thomas and Sillen's *Racism and Psychiatry* (1972).

[37]A program which sponsored American black and white youths to work and visit in selected African nations during summers.

[38]This is similar to the problem raised by Pierce (1968) in his discussion of the increasing demands upon the black American which will emerge upon the diminishing of the deprivations.

[39]See Pettigrew, T., "Negro American Personality: Why isn't More Known?" *op cit.* The recent (since 1970) contributions here are reviewed in our concluding chapters.

account for the complexity. Although sorely needed, it does not seem that a "psychological theory" of the black American is yet available. But is a new framework or specific social psychology role theory required? Is it possible that a framework as complex and general as psychoanalytic ego psychology will be able to deal with much of the data of black American behavior and development? Such a thesis is at least worthy of consideration before building new conceptual mazes. Exploration of this possibility is the course taken in the research in black adolescence presented in this monograph.[40]

As will be seen, there have been several comparative studies of American black and white populations. Supposed black-white differences in personality development, character, abilities, and psychopathology have fascinated and puzzled investigators. Black and white populations are again contrasted here. But this time the comparison is along very specific lines which have not been used in the prior comparative studies. To be analyzed is an interface between psychoanalytic considerations of individual development and the sociocultural matrix. The interface chosen is an aspect of adolescence in which the individual must synthesize past identifications within himself, while also in some way finding a significant place for himself within the structure of his community. It is this aspect of adolescence where *consolidation* and *continuity* are crucial issues of maturation.

Some psychologically oriented studies of current desegregation efforts have been reported by Coles (1963,1965a, 1965b, 1965c, 1967). His studies consist largely of impressionistic observations and informal, sometimes intensive, interviews with both white and black children and their families. Most welcome is his inclusion of whites, who become interesting in themselves as well as a needed "control" group.

When we more narrowly seek pertinent literature as being *only* that dealing with black psychological development, in particular adolescence, we suddenly find a sharp decline in available research. The first study specifically of black adolescents was made in 1938 by Dollard and Davis.[41] Two hundred and seventy-seven upper, middle, and lower class New Orleans teenagers were interviewed and 76 chosen as "intensive cases." The latter and their families were given multiple

[40]In our last chaper we return to discuss the question of a "black psychology".
[41]See Dollard and Davis, *Children of Bondage* (1941). A more recent study is that by Pierce (1968). This was found after the writing. It is primarily a series of speculations and concerns about the effect of the now lessening deprivations and degradations. As the black "underprivileged" now become privileged, what will the impact be for black adolescents? It will be recalled that Harold Isaacs raises a similar more general form of this question.

interviews and psychological tests. The results of the study at that time were interpreted as primarily consequences of the differing social classes: caste distinctions being assumed to influence all subjects in an essentially uniform manner.

The same subjects were re-studied 20 years later by Rohrer and his associates (1960). In the re-study both social class and "self-hate" as single factor explanations of the psychological development of the subjects were rejected. Instead, the subjects were found to fall into five major groups, which were based on primary role identifications, "patterns of cultural identifications, instituted in family life and in the manner of training children." Viewing their subjects primarily in terms of orientation to "one cultural nucleus" be it "matriarchy," "gang," "family," "middle class," or "marginal," is defended by the authors as demanded by the great variations they found when comparing individual development. The authors readily admit, however, that the analysis is incomplete as it stands.

We are left with a clinically rich study of the social and psychological changes manifest by a selected, heterogeneous group of New Orleans blacks. Conclusions drawn are modest and cautious. Several of Rohrer's explanatory concepts are similar to the ones of this study. For together with "identification" with cultural orientations, the term "identity" is adopted in order to distinguish the intrapsychic from the cultural dimensions of subjects' experiences. Identity in Rohrer's terms denotes:

> . . . certain comprehensive gains that the individual at the end
> of adolescence must have derived from his pre-adult expe-
> riences in order to be ready for the task of adulthood.[42]

But this term is a minor one in the subsequent discussions of results. And when used it is taken as roughly synonymous with self-image. The limited generalizations and mixture of social psychology, anthropology and psychoanalytic concepts is in part a consequence of the study's "team" approach of anthropologists, psychologists, sociologists, and psychoanalysts. Such a group obviously was productive of a multifold perspective on the data. Yet one suspects that it also hindered the use of any single theory as well as the formulation of hypotheses from the data. There are many points of convergence between Rohrer's work and the study of New Haven adolescents reported here. The most important difference will lie in our adher-

---

[42]Rohrer, et al.

ence to different theoretical frameworks for the analysis and interpretation of the investigation.

Kardiner and Ovessey (1952) also intensively studied black adults and adolescents. They differ from Rohrer in both method and interpretation. Using modified psychoanalytic interviews, they examined 25 New York blacks of varying age, sex, and class. On the basis of their detailed case histories they conclude that there are conscious and unconscious trends of self-hatred and "identification with the white" in all of their subjects. In evaluating their conclusions it is important to consider the fact that half of their subjects were also explicitly "patients," receiving free psychotherapy instead of cash for being informants. Moreover, Rohrer in reviewing Kardiner's case records, claims that there is evidence of conscious self-hatred in only 7 subjects, five of whom were patients.[43] The existence of self-hatred in all American blacks obviously cannot be convincingly argued from numerically limited and then probably psychopathologically biased samples. Only through observations of many black subjects from various health, geographic, and class backgrounds can we answer this question, or the more general one of generic responses and tendencies among black Americans.

Studies of black developmental problems and adult psychopathology are of course valuable in suggesting specific conflicts and trends which may characterize certain groups of, or possibly all, black Americans.[44] There are some findings common to all of these studies: the black subject or patient is most often described as having low self-esteem, marked unconscious aggression and hostility, and a degraded self-image.[45] Sources of data range from free association interviews to the Thompson Picture Arrangement Test (PAT). Karon's (1958) study is distinctive in using the latter "standardized" projective instrument, and in his emphasis on the differences between Northern and Southern blacks. The latter differentiated themselves from their Northern counterpart on the basis of greater concern with aggression, weak and labile affect, and conflicting work motivation.[46]

Probably the most difficult productions to study systematically are the expression of black literary figures. These writings are nonetheless

[43]Rohrer, E., op cit., p. 72.

[44]See Rose (1956), Dai (1955), Powdermaker (1955), Sclare (1953), Karon (1958).

[45]As we elaborate later (Chapters 8 & 9), these conclusions have been most seriously challenged by new findings and interpretations.

[46]Black Rage is a more recent study of American blacks with a dominant clinical focus. Through much rich interview material and more general social commentary several of the issues found in the above writings in Tally's Corner, are elaborated upon. See Grier and Cobb, op cit.

of inestimable importance for insights which may not be gained through other means. Popular writers such as Silberman (1963) have surveyed many of the literary works. White (1947) undertook a value analysis of *Black Boy*; Bone (1958) studied the works of several black writers. In general, however, these literary works have not been seriously analyzed. The essays, books, and plays are a rich and largely unmined source of important data and understandings about the American black.[47]

A literary form of special interest to the study of identity formation is the autobiography. At its best, the autobiography provides the chance to trace individual trends of development, their internal consistencies, their thema, and the articulation of these trends with social and cultural patterns. A number of black autobiographical works have been presented.[48]

Clearly, these autobiographical works represent a growing "file" of case studies whose careful analysis can only contribute needed data about black personality development and experience. In *Manchild in the Promised Land* Claude Brown richly describes a childhood and adolescence spent in a ghetto, followed by prison. He notes a series of changes in his life style and orientation, touching occasionally upon possible determinants of the shifts. It is likely that further analysis of the material would lead to other formulations about his development. Such a possibility exists for each of the autobiographies cited.

This comparative study of black identity formation has been informed in many ways by the essays, novels, and autobiographies as well as by the more "objective" empirical research. There is a large and ever-expanding literature on black Americans. From this literature, a series of significant contributions were selected and discussed in this chapter. Generally omitted from more detailed review were the fictional and autobiographical works. Such an omission is in no way

---

[47]Best known of the authors in this genre is James Baldwin. His prolific works are filled with themes of protest, injustice, emasculation and degradation. These themes are developed in all the novels as is, to a more variable extent, the explicit problem of being black. See *Go Tell It On the Mountain, Giovanni's Room, Another Country*, as well as the collected essays in *Nobody Knows My Name*, and more recently, *The Fire Next Time*. MacInnes (1963), is an interesting summary and discussion of Baldwin's writings. Other important contemporary black writers and writers on blacks include Wright (1945), Redding (1951), Ellison (1952), Miller (1959), Hansberry (1959), Jones (1967), and Fanon (1962).

[48]These include Malcolm X, *Autobiography of Malcolm X* (1967); James Baldwin, *The Fire Next Time* (1963); Eldredge Cleaver, *Soul on Ice* (1968); Saunders Redding, *On Being Negro in America* (1951); and Claude Brown, *Manchild in the Promised Land* (1965).

related to their importance. It does testify to the difficulty of treating these productions. One approach could be analyses of each work, the results of which might then generate new hypotheses and formulations. Alternatively, systems of content analysis might be directly applied to each specific work as means of investigating particular hypotheses and issues.

Within the group of empirical studies, there are three which can be said to have immediate bearing on this research: *Children of Bondage*, *The Eighth Generation*, and *The Mark of Oppression*. It is only in these efforts that black adolescents are a focus of study as well as discussion. Although the present study differs from these in terms of both theoretical bases and techniques, they nonetheless provide valuable comparative data.

At this point, then, we have an overview of the germane theoretical and empirical background to the study of black identity formation. With this perspective in mind, we can proceed to full examination of the patterns of identity formation which we studied in the New Haven adolescents. The remaining chapters will deal with all aspects of this investigation, moving from the study itself to speculative models for understanding the discovered patterns of identity formation, and finally to further considerations in light of recent contributions.

CHAPTER 2

The Measurement of Identity Formation

PRELIMINARY CONSIDERATIONS

A definition of identity formation taken from one of Erikson's papers was quoted in Chapter 1. The notion of identity formation as an "evolving configuration" of intrapsychic and psychosocial variables appears in various forms throughout Erikson's writings, be they clinical, historical, or theoretical. Generally, these formulations are well suited for intuitive understandings of the case studies and essays and even at times for theoretical clarifications. Yet to construct means of objective measurement, which can then be used in empirical studies of identity development, these descriptive statements must be transformed into a more precise composite definition which can be used by clinical and nonclinical investigators. The translation of descriptions of identity into a more "operational" statement is discussed and proposed in this chapter.[1]

The configuration or "integration" of components that Erikson calls "identity" is not restricted to adolescence. Development of identity has been progressing through all prior psychosocial stages. In adolescence, however, the process becomes problematic. It becomes increasingly conflictful for a variety of intrapsychic and psychosocial reasons, foremost among them being: the onset of puberty, bringing with it renewed oedipal conflicts and the demands of increased libidinal drives; cognitive changes such as increased awareness of irreversibility; and new societal demands in terms of work, sexual commitment, and ideological commitment. This "primacy" of ego identity is manifested by the individual's conscious and unconscious preoccupations with consistency, personal and social significance, commitment, and irreversibility. Most inclusive of these concerns is that of personal continuity: a sometimes intense sense of physical, ideological, social, and emotional

---

[1] A very different strategy for operationalizing "identity formation" is presented by Marcia (1980), as discussed in Chapter One.

change. More concretely, a heightened awareness of continuity and "wholeness" appears through themes of body image, sexual definition, social roles, values and ideals, and interpersonal intimacy.

"Successful" resolution of the task of identity formation is said to be heralded by the individual's gain in sense of direction, meaning, and overall personal and social coherence. Concomitantly, there is a decline in his preoccupation with synthesis, continuity, and the other themes noted above. One other sign of successful resolution, predicted by Erikson, is the emergence of the capacity for and interest in sustained heterosexual intimacy: the critical psychosocial task of the next maturational stage.

However, continued or intensifying conflict over problems of personal continuity, unremitting preoccupation with identity themes, or the inability to attain interpersonal intimacy are all signs of nonresolution and "identity diffusion". Another pattern of nonresolution is that of identity foreclosure. This has surface resemblance to successful consolidation. There is a diminution in all of the above conflict areas. But this is where the similarity ends. Rather than synthesis and richer definition, the result is impoverishment and restriction. The conflicts and confusions have been lessened by avoidance. The individual's sense of direction and self is solidly fixed, as it is in response to a withdrawal from conflict areas, a rigid closing off of possibilities. A sub-type of identity foreclosure is "negative identity". In identity foreclosure (or negative identity) the possibility of intimacy is highly unlikely.

AN OPERATIONAL DEFINITION

Identity formation is a construct that refers to a particular phase in ego development. The nature of this step in ego development is closely dependent upon social and cultural contexts. These factors must be considered in fully analyzing any given case of identity formation. When dealing with the problem of the means for objectively specifying and measuring identity formation in a given individual or group of individuals, however, such contextual matters can and must be set aside. They become relevant following determination of identity formation itself. It is then that the questions of changes, relationships to other ego identity patterns, and deviations from predicted development can be raised. Answers to these and similar problems require that social and cultural processes be taken into account in addition to the psychological ones.

The advantages of formulating a workable index for identification and measurement of identity formation are more than obvious. The host of hypotheses implied in the clinical studies can gradually be tested by investigations on many and varied populations. The validity of the inferences derived from this clinical construct can be evaluated and, in addition, further theoretical implications clarified and discovered.

In this section, the operational definition of identity formation is first detailed. After the full statement of the definition itself, each of its components receives further discussion.

Identity formation designates specific processes involving an individuals self-images, namely: (1) their structural integration, and (2) their temporal stability. Changes in either of these processes is indicative of specific types of identity formation.[2]

It is necessary to examine each term of this definition to further clarify its empirical meaning.[3]

Self-Image

The term "self-image" refers to those concepts, conscious, preconscious, and unconscious, by which an individual characterizes himself. Identity has often been described as referring to "consolidation" or "integration" of multiple components of ego function such as capabilities, significant defenses, identifications, and ideals. The term "self-image" includes these other components. In other words, self-image is a higher level of abstraction; each self-image may be further analyzed into the above elements. To understand a given individual fully, to come to grips with the significance of his identity as part of his overall personality development, it is necessary to investigate the more tradi-

---

[2]This is a *formal* definition, specifying those general qualities—criteria—that indicate the presence, absence, or current state of the ego development called identity formation. There are also *content* aspects. Certain content variables, psychosocial issues and themes, are postulated as being related specifically to the development of ego identity in adolescence. Such issues include preoccupation with body image, sexual prowess, role definition, values and ideals, and conflicts over interpersonal intimacy. They are "markers," reflecting the psychosocial events associated with identity at this point in the life cycle.

Since ego identity develops throughout the life cycle, it is therefore possible that different issues or "themes" specifically linked to it are prominent at other times. This possibility has never to our knowledge been investigated. Furthermore, not only are the issues "phase specific," but they also may be culturally and socioeconomically conditioned both in terms of quality and intensity. This latter proposition is suggested in following the development of the two boys discussed in Chapter 6. Much of the data there is related to ethnic and adolescent factors in the identity development of two subjects.

[3]Reflections about this, and the subsequent operational definitions, were greatly facilitated through discussions with Dr. Ernst Prelinger; particularly in relation to his use of the Q-sort.

tional intrapsychic ego functions, as well as those elements included in id, super ego, and ego ideal constructs. However, the notion of identity formation qua identity formation does not *directly* concern itself with anything but self-images and in particular the two basic aspects as noted.[4] There are probably many techniques for objectively determining the array of a person's self-images. Obviously, psychoanalysis and psychotherapy are means for this. A more immediately objective and empirically verifiable one is by means of the Q-sort, which is the method used in our study. Through this technique, an individual sorts—in the sense of rating—given groups of statements about himself under specific conditions, such as his self-image now, or his self-image last year, or in 10 years. The statements can be a statement set used for all subjects, or they can be those taken from the subject's various self-descriptions. The latter method probably assures that the elicited sorts most closely correspond with the subject's verbally expressed self-images.

Structural Integration

The many clinical definitions by Erikson and others emphasize the "consolidation" processes of identity formation. There is a coherence of elements, a growing "organic" whole. Translating the valuable intuitive impression to a measurable function gives the concept of "structural integration." By this is meant the intercorrelations of an individual's self-images at any given time as measured by the Q-sort. These are the *intrayear correlations*. Structural integration is measured by the average of all these intercorrelations at any specific point in time. When dealing with more than two self-images, as is almost always the case, one can also speak of those more-or-less closely integrated images suggesting a hierarchy of importance. Finally, if the same subject is seen over any period of time, he may be retested for changes in this identity formation process.

Temporal Stability

Development of identity is also reflected by changes and consistencies of self-images. When speaking of identity formation, the concept of *continuity* is frequently used. One important meaning of this is constancy of self-images: the extent to which present self-images resemble those the individual held in the past. In these clinical descriptions an

---

[4]Erikson (1968) notes that identity formation can have a "self aspect" and an "ego aspect." He notes that "self-identity" refers to the integration of the individual's self and role images. In these terms, then, this is a formal operational definition of the "self aspect" of identity formation.

"increasing sense of continuity." is taken as an important indication of identity formation; a decrease in this sense suggests the opposite. Again, this intuitive impression can be operationally defined by the Q-sort. The change or constancy of any specific self-image can be measured by obtaining the correlation between the sorts made under the same conditions at any two different times. Such determinations are *interyear correlations*. This can of course be done with any self-image, providing the conditions of sorting and number of statements are held constant. For any given time interval, the temporal stability process of an individual's identity formation is determined by the average of all interyear correlations.

This process can obviously only be measured over time, supporting the proposition that identity formation must be investigated by a longitudinal approach. It is conceivable that some observations and indices relevant to identity formation can be obtained by a single study. As more is learned of identity development, these partial determinations may be useful in assessing states of identity formation. In Chapter 9 we further discuss the topic of longitudinal studies in this area.

VARIATIONS IN IDENTITY FORMATION

Using the definition presented above, the variants of identity formation observed by Erikson can be recast into operational statements. Individuals can then be typed into these clinically useful categories on the basis of clear, publicly agreed upon criteria. There is obviously no reason why one must adhere to these clinical types; the study of other populations using the Q-sort and similar measurements of self-image interaction may suggest even more useful categories or, more probably, finer distinctions within these. Erikson's classifications are employed here, however, primarily because of their prior utility in clinical discussions and the amount of pertinent clinical data that he and others have collected.

Progressive Identity Formation

Throughout childhood and adolescence, ego identity develops, with an acceleration of this process occurring in adolescence. Less specific remarks have been directed toward the period after adolescence. It is implied, however, that the process slowly continues, reaching a climax in old age, the "wisdom" stage of the life cycle.[5] We may describe an individual as manifesting progression in identity formation when *both the structural integration and the temporal stability of his self-*

[5]See E. Erikson, "Growth and Crises of the 'Healthy' Personality," *Psychological Issues*, 1, p. 98 (1959) and Erikson (1968).

images are simultaneously increasing over any given period of time, though not necessarily at the same rate.[6]

Identity Diffusion[7]

This clinical type designates those cases in which there is failure to achieve the integration and continuity of self-images. The category is a broad one and probably includes several subtypes, for there are conceivably multiple etiologies underlying this outcome of identity formation. Such a state may be present at any stage of the life cycle. However, it is theoretically most likely to be manifest at adolescence. Again, only through further empirical investigation can this hypothesis of stage specificity be studied. For example, does identity diffusion occur during latency or unexpectedly in adulthood and also hinder continued psychosocial development?

Operationally, an individual is said to be in a state of identity diffusion when *both processes of his self-images show a repeated decline in magnitude over any given period of time.* Again, these changes need not be at the same rate. In fact, differing rates may be one sign of variant kinds of identity diffusion. However, of the two processes, structural integration is the most critical here in defining the state of diffusion. Another possibility in this category is a form of *"attenuated diffusion"* in which either (1) structural integration decreases and temporal stability remains constant, or (2) a milder type in which temporal stability decreases and structural integration remains constant. It is possible that these represent early forms of flagrant identity diffusion in which reversibility is more likely.

Identity Foreclosure

This variant in identity formation superficially resembles progressive identity development. There seems to be present a sense of integration, "purpose," stability, and a diminution in subjective confusion concerning these matters. However, the stability and purpose are the result of an avoidance of alternatives, of a certain restrictiveness which eliminates ambiguities. What appears to be the outcome of a successful progression in identity formation is actually an impoverished, limited self-definition and sense of continuity. The strains and confusions inherent in the syntheses necessary for adolescent identity formation have been

[6]Formal symbolic statements describing this development, as well as the variants, are given in Appendix C.

[7]In his discussions, Erikson now refers to this variant as "identity confusion." We retain the previous, more *familiar* term here, however; see Erikson (1968), p. 131.

bypassed, as the developing individual has settled upon a certain identification or set of identifications as forever characterizing himself in all ways.[8] Identity foreclosure is thus an interruption in the process of identity formation. It is a premature *fixing* of one's self-images, thereby interfering with one's development of other potentials and possibilities for self-definition. An individual does not emerge as "all he could be."[9]

Operationally, a person is said to manifest identity foreclosure when *either or both the structural integration and temporal stability processes of his self-images remain stable*, or only temporal stability shows continued increase while structural integration is unchanging.[10]

Negative Identity

In this identity formation variant, configuration of self-images is fixed upon those identifications and roles that have been presented to the individual as most undesirable. The individual sense of identity is based on the repudiated, the scorned. With its emphasis on those historically repudiated and rejected identifications, this variant represents a premature closing off of new syntheses of identifications, of new identity configurations. As such, it is a type of identity foreclosure. Although theoretically it has different developmental roots than identity foreclosure,[11] the structural pattern is the same: abortive identity formation, premature fixation of self-images, thereby halting further evolution of self-definition.

It follows that the operational definition of negative identity is the same as that of identity foreclosure. Structural integration and temporal continuity are unchanging, remaining at the same level year after year. Or if there is any change, it is in the temporal continuity process that increases its level of activity and thus reflects even greater similarity

---

[8]This is not to imply that the avoidance of these issues is the determinant of identity foreclosure. It is but one consequence. As can be seen in Chapters 6 through 9, the determinants of this variant are indeed a complex matter.

[9]E. Erikson, "Growth and Crises of the 'Healthy' Personality," *Psychological Issues*, 1, 1959, p. 87. In that section Erikson gives examples of "good little worker" or "good little helper" as illustrating such premature fixing.

[10]It is conceivable that this structural integration could be at a very low level of correlation and thus be more representative of an interrupted identity diffusion. The definition of foreclosure may therefore require a numerical specification such as: "both structural integration and temporal stability must be .40 or greater."

[11]Theoretical aspects of the genesis of both *identity foreclosure* and *negative identity* are discussed in Chapters 8 and 9.

of self-images over time. To differentiate negative identity as a type of foreclosure, the actual content of the self-images must be examined both with respect to one another and even more importantly, with respect to knowledge of the individual's own history.

Psychosocial Moratorium

An individual is described as being in a psychosocial moratorium when he is "finding himself," experimenting with varied roles, new self-images and future plans, at all costs remaining uncommitted to any particular alternatives of self-definitions. At the same time he is *not* tending toward, or experiencing, a type of identity diffusion. This is, of course, the *antithesis* of foreclosure. Rather than rigid sameness, the content and patterns of self-images show continual variation. The key concept here is "openness," noncommitment. No irreversible decisions or plans are made. A partly conscious, partly unconscious attempt is made to ensure maximal flexibility and diversity before the further elimination of any possibilities, alternatives, or actions inherent in all stages of identity formation. Operationally, an individual is in the period of *psychosocial moratorium when the temporal stability and structural integration of his self-images shows significant fluctuations (increasing and decreasing) over a given period of time.* Consistent with the tendency running counter to diffusion, the structural integration feature shows less fluctuation than that of temporal stability, particularly in terms of any decrease in value.

Implicit in all of the preceding definitions is the notion that identity formation is a developmental process. The criteria specified by each operational statement require *time-based comparisons* of correlational levels. According to this position, determination of any type of identity formation rests upon observations taken at several points in time. It follows that there are no absolute criteria with which one can define identity formation variants. Even more specifically, observations from a single point in time cannot by themselves adequately characterize this *developmental* process. It is a patterning over time of particular variables that serves to define the type of identity formation taking place.[12] In interpreting the data presented in Chapter 4, we on several occasions need to recall this key notion of time patterning.

The operational definitions of identity formation variants and their clinical counterparts, are summarized verbally and graphically in Tables 1 and 2. Having reviewed our operational definitions in some detail, we now turn to look more closely at our sample and specific measures.

---

[12]It is this insistence on multiple observations over time that most separates this study from the few other empirical studies of identity development. Keniston (1959), Dignan (1965), Marcia (1980), and the several students who use Marcia's measures all use cross-sectional designs.

Table 1. Identity Formation Variants

| Variant | Definition | |
	Clinical	Operational
Progressive identity formation	Continual increase in synthesis of identifications, personal and social continuity. "An evolving configuration of constitutional givens, idiosyncratic libidinal needs . . . consistent roles."	Consistent increase in both structural integration and temporal continuity.
Identity diffusion	Continual decline in synthesis of identifications and ego functions; decreasing sense of wholeness and continuity with self and community. Fragmentation.	Progressive decline in structural integration and temporal stability; more severe form has greater decline of structural integration.
Identity foreclosure	Rigidity in self-definition. Lack of any change in synthesis of identifications or other synthetic processes. Characterized by premature aborting of identity development. At first glance resembles "successful" identity formation.	Little or no change in temporal stability or structural integration. If there is any marked change, it is in the direction of increasing temporal stability, in face of unchanging structural integration.
Negative identity	Premature self-definition based on *repudiated*, scorned identifications and roles. Commitment to what is personally despised.	Same as identity foreclosure, of which it is a subtype. Examination of *content* of self-images essential to distinguish it as a subtype of foreclosure.
Psychosocial moratorium	"Experimental" state; no firm commitments made. "Trying on" of roles and integrations characterized by flexibility, flux, but *not* disintegration.	Fluctuation, swings, in both directions of temporal stability and structural integration.

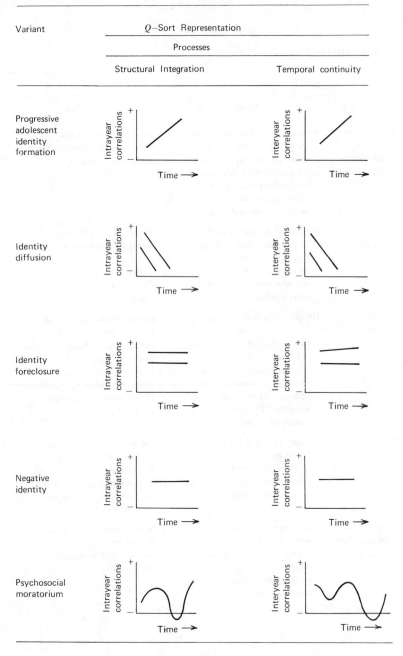

Table 2. Identity Formation Variants

CHAPTER 3

The Structure of the Study

THE SAMPLE AND TIMING OF THE INVESTIGATION

The names of possible subjects were obtained from New Haven junior and senior high school guidance counselors and assistant principals. These school officials were requested to choose students who had the following characteristics: male, entering sophomore year of high school, from a working class family[1] and neither delinquent, predelinquent, nor "college-bound." The subjects were then seen individually and accepted for the study on the basis of the above criteria and their motivation to participate. A total of 23 black and white boys were initially interviewed and tested. From this group five were not seen again after the first year; other "dropouts" occurred in later years. Table 1 summarizes characteristics of all subjects who participated in the study.

In brief, the sample consisted of 11 black, 11 white, and 1 Puerto Rican boy. Their chronological ages at the start of the project varied between 14 and 16 years old; there was no significant difference between the average age for either major group. Socioeconomically, 22 of the 23 boys were from Class 4 or 5 families, as determined by the Hollingshead (1957) social class index.[2] Eight of the 11 blacks were Class 5 ("lower-lower class"), while 4 of the 11 whites were from this class.[3] Of interest in

[1]All subjects who on the basis of the Hollingshead (1957) index were from a Class-4 or -5 family, were considered to be "working class".

[2]This index is intended as an objective, easily applicable instrument for determining the positions individuals or households occupy in the status structure of our society. It is based on the head of the household's occupation and amount of formal education. Rationale for the index and further description of its application can be found in Hollingshead (1957).

[3]It is possible that the Hollingshead index is not applicable to nonwhite groups without some modification in either scaling or factor weighing. Brown (1955) and Glenn (1963), among others, have suggested that the symbolic prestige value for both

Table 1. The Sample

Subject	Age[a]	Race[b]	Socio-economic Class[c]	Number of Years in Study[d]	Head of Family
FM	15.0	B	4	3	Father
GR	14.5	B	5	3	Father
LT	15.0	B	5	3	Father
OS	14.0	B	5	1	Father
FL	14.0	B	5	2	Mother
MD	16.0	B	5	3	Mother
TM	15.0	B	5	3	Mother
BA	16.0	B	5	3	Mother
KE	14.5	B	5	3	Father
JE	15.0	B	4	2+	Father
KB	16.0	B	4	1	Father
HN	15.0	PR	5	3	Father
AL	15.0	W	4	2	Father
NS	15.0	W	5	3	Father
BT	14.0	W	4	3	Father
ND	16.0	W	5	1+	Father
VR	15.0	W	4	3	Mother
JK	16.0	W	4	3	Father
JR	15.0	W	5	3	Father
IQ	16.0	W	4	1	Father
CS[e]	15.0	W	3	1[c]	Father
LE	15.0	W	4	1	Father
DM	15.0	W	5	1	Father

[a]Age at start of study, to closest half-year.
[b]B, black; PR, Puerto Rican; W, white.
[c]According to Hollingshead index; see footnote 2.
[d]Three years represents completion of study.
[e]Dropped from the study because of both his social class and college goals.

regard to attrition is that all black dropouts (from the study) and 3 out of the 5 white dropouts were from Class 4 families ("upper-lower"). This is but one of several interesting features that characterized the subjects who left the study. These subjects are described in greater detail in Appendix B.

education and occupation may be very different among blacks when compared with white populations. Roughly, the same job or education is seen by blacks as "worth more" than it would be in a similar group of whites. At the time of this study, no systematic modification based on ethnic or racial differences had been derived for the Hollingshead index, or for any other social class index. Obviously, this is sorely needed. For the data here, it is likely that we have a "false low" for the black families. Probably then, the black and white boys are from approximately the same socioeconomic class.

Geographically, the subjects came from many areas of the city; from Rockview and Elm Haven Project, Dixwell Avenue, Wooster Square, Fair Haven, and central New Haven. Explicit effort was expended to avoid choosing the sample only from "ghetto" areas. Equal numbers of subjects came from each of the two public high schools; one subject attended a state trade school. Other features of the sample, such as family composition, geographic mobility, and school performance are presented in Appendix B.

The sample, then, is a highly specific one. It is, to begin with, all male. In addition, it is limited to members of lower socioeconomic classes. These constraints were self-imposed for very specific reasons. One purpose of the study was to investigate *non*-middle-class youth in terms of identity formation, hence the limitation to lower-class populations. It should be of much interest to make comparisons across class as well as ethnic lines using the techniques of this study.

A second constraint of the sample is that of sex. There is no reason to assume that identity formation is similar for males and females. However, rather than study possible intersexual differences in these developmental processes in addition to possible interracial ones, the choice was made to limit the sample to males and thereby study one specific contrast within an otherwise generally homogeneous group. In other words, sex and social class were held constant in the group of subjects, while race was selected as the contrast characteristic.

All subjects were paid at the rate of one and one-half dollars per hour. Initially, and then twice each year until graduation from high school, each boy was seen for a combination of interviews and tests.[4] The schedule of interviews, and tests of this first year and all subsequent ones is given in Table 2.

Generally, interviewing and testing were carried out individually at neighborhood schools. When this arrangement was not feasible, an interviewing room with built-in tape recording equipment at Yale Medical School was used. Whether seen in the latter facility or at the schools, all subjects were informed that the interviews were being recorded and that the recordings would be kept anonymous.

THE TECHNIQUES OF INVESTIGATION
The Interviews

All interviews were tape recorded and later transcribed. Inasmuch as possible these sessions were unstructured, the interviews being guided

[4]These tests included the TAT, Sentence Completion, Draw-a-Person, Prelinger-Otnow Intimacy Test, and Q-sort. All tests, except for the Q-sort and Draw-a-Person were administered each year, in the spring.

Table 2. Schedule of Procedures

Year	Dates	Interviews	Tests[a]
First	August— September	Initial interview Work Heroes and myths Body image and intimacy	Sentence completion Draw-a-Person TAT Intimacy Test Q-sort
	June	Interim interview	Draw-a-Person Q-sort
Second	October	Interim interview	
	June	Interim interview	Draw-a-Person Q-sort
Third	October	Interim interview	
	Spring	Interim interview and repeat topics of all initial interviews	Repeat all of initial tests, including Q-sort
Fourth+		Follow-up interviews	Follow-up Draw-a-Person and Q-sort

[a]These tests were given in the order described here. However, it is the results of the Q-sort that are presented in detail in the text. The other results are not reported here. Description of the other procedures can be found in Machover (1949) (Draw-a-Person), Dorris *et al.* (1954) (sentence completion), and Otnow (1962) (intimacy test).

only by a list of topics to be covered. These topics were largely explored through open-ended questions which were introduced during the initial set of interviews and all later "interim" interviews. Thus while the interviews were "free flowing," inclusion of a standard set of topics in each set of interviews was assured. The interim sesssions were devoted to reviewing recent experiences and events as well as new plans, wishes, ideas, conflicts, and decisions that had occurred since the last interview. Historical matters were discussed at length where appropriate during some of these sessions. Usually, the course and content of the interim interview was determined by the subject as he brought the interviewer up to date and chose to pursue related new issues.

The interviews and tests were all conducted by the same interviewer[5] who was first identified to the subjects as a medical student interested in studying "how teenagers change." It soon was apparent that this information did not completely clarify who he was and what his

[5]The interviewer and tester throughout the study was the senior author, (STH).

purposes were. Repeatedly, subjects showed confusion and distortions regarding his professional status and some anxiety about his motives.

As might be anticipated, the most extensive clinical data of the study was generated from these interviews. The material was studied in three ways: by impressionistic analysis for prominent themes and individual patterns of development; by application of objective scales and codes by independent judges; and, finally, as a means of generating statements for the yearly Q-sorts. The Q-sort and thematic analysis are those procedures most directly related to the results reported in this monograph. It is therefore important that we now examine the nature of this valuable instrument.

The Q-Sort[6]

From each interview all explicit self-descriptive statements made by the subject were abstracted. These statements concerned his attitudes, plans, wishes, fantasies, feelings, judgments, actions, or thought. Each of the statements was then phrased in the simplest possible form, that is, as a positive statement in the present tense and without qualifiers or conditional clauses. The statement was then placed on an individual three-by-five card. The complete pack of such cards became the subject's "deck."[7]

Each year, within one or two weeks of the set of interviews, the subject was presented with his deck of statements. He was told that this numerically coded set of cards was based on what we had discussed during our several conversations. Ten four-by-six cards numbered individually with large numbers from 0 to 9 were laid out on the table in front of him, in ascending order. He was then requested to judge each statement in terms of how well it described him, placing the most accurate descriptions at the 9-position, labeled "most important," and the least accurate ones at the 0-position,

[6]The specific use of the Q-sort in this study is an elaboration of its application by Prelinger, as described in his NIMH Grant Proposal M-3642. The help of Dr. Prelinger in developing and using the technique for this study was considerable and is gratefully acknowledged.

The Q-sort itself, and the more general methodology from which it emerges, has been summarized and most thoroughly discussed by Stephenson (1953). This work provides a basic introduction to the method. The orientation of Stephenson, originator of the Q-technique and methodology, is aptly characterized by Brown (1968): "Stephenson has been most concerned from the start with providing an objective approach to the problems of subjectivity" (p. 589). Brown's (1968) comprehensive bibliographic review of this entire area gives clear evidence of the extensive and varied applications that have been made of this technique in the past 30 years.

[7]Examples of these statements are given in Appendix D.

labeled "least important." He could place those statements that he felt did not belong at either extreme at any of the intermediate positions (1-8), depending on the extreme to which they were closer. Or, statements could be judged as neutral and placed on the number 4- or 5-position. Hence the subject was required to sort his statements on a 10-point scale of importance; the only distributional requirements being the use of each of the two extreme categories at least twice.

Using the same deck or its exact duplicate, the subject then sorted the cards seven additional times, once for each of seven other self-images. Always in the same order, the self-image instructions were introduced one at a time for seven more sorts. In summary, then, the subject was told to arrange the deck of statements according to how well they described him for the following possibilities:

1. How am I now
2. How I would be if I were a perfect son to my mother
3. How I appear in the eyes of my friends
4. How I will be in ten years
5. How I would be if I were a perfect son to my father
6. How I appear in the eyes of other people
7. How I was at the beginning of junior high school
8. How I would be if everything worked out exactly the way I want it to; how I would be if all my dreams came true.

This procedure was first performed following the initial set of interviews, in the fall of sophomore year of high school. Thereafter, it was repeated in June of every year using the subject's original deck of cards, to which had been added all new self-descriptions made in the individual interviews. Thus the deck "grew" annually. No statements were ever deleted. For all subjects who completed the study, a total of four complete sets of sorts were obtained.

Two different product-moment correlations can be derived from these data.[8] First, the correlations of each year's self-images with one another can be calculated. A total of 28 such correlations were determined each year for each subject. These correlations are *intrayear* correlations and represent the degree to which the subject's

---

[8]These correlations, as well as all average correlations, were computed through the IBM 7040/7049 system. Additional information from these sorts includes the content of those statements that are consistently rated at the extremes for individual subjects. (The results of such an analysis will be presented at a later date.) Both sets of analyses were greatly facilitated by the use of the Yale Computer Center and the able assistance of Roger Bakeman, programmer at the Center.

self-images resemble or differ from one another in any given year; they are estimates of the "fit" of the images.

The second kind of correlation is related to change over time. The change in a self-image was determined by comparing the arrangement of statements it elicited in any two different years. Suppose, for instance, that we want to compare the individual's description of his current self (Sort 1) in the first year of the study with the same self-image two years later. To accomplish this comparison quantitatively, the correlation between the sort of his cards for this image in Year 1 and in Year 3 would be calculated. This type of correlation is an *interyear* correlation; it represents the exact quantitative determination of the change or constancy of any given self-image.

Through our Q-sort technique and these forms of quantitative analysis, the identity formation of the black and white subjects was studied. We now turn to examine the longitudinal data which were analyzed through this method and the linked operational definitions of identity formation.

CHAPTER 4

Black and White Variations in Identity Formation

INTRODUCTION

A group of black and white adolescents were individually followed from the start to the completion of high school.[1] In this chapter the results of the quantitative Q-sort studies are presented and subsequently analyzed in terms of identity formation patterns.

Four sets of self-images are discussed: parental self-images, personal time self-images, fantasy self-images, and current self-images. Relations within and between these sets are reviewed by examining annual interyear and intrayear correlations and changes in these values over time. Following the survey of these self-images, we consider the specific indices of identity formation: the average interyear and intrayear correlations.

By design, this chapter is a summary of numerous correlations from each of the subjects. The subjects sorted large decks of cards annually for four years. On these cards were specific self-descriptions of many kinds,[2] which each subject had given in the prior interviews. The cards were arranged—sorted—by the subject in terms of how accurately they described each of eight self-images. Hence, every year a subject resorted his set of cards to describe these eight self-images.[3]

---

[1]This group consisted of 22 subjects at the outset of the study. With varying attrition rates, the number varied over the subsequent years. In the significance tables included in Appendix A, the exact number and racial composition of subjects per year interval are given.

[2]Chapter 3 gives general descriptions of these statements. Illustrations of specific statements and how they were arranged are presented in Appendix D.

[3]The specific self-images are detailed in Chapter 3 and also in later sections of this chapter. The way in which this Q-sort was used here, and most specifically the idea of

The degree of similarity between the Q-sorts for any two self-images at one point in time is an *intrayear* correlation. There were 28 such correlations every year for every subject. The average of these 28 correlations describes the process of structural integration at a given point in time. Structural integration is one of the two basic processes of identity formation described in Chapter 2.

The degree of similarity between the features of a single self-image in two different years is an *interyear correlation*. Each subject had eight such correlations for any given time interval. The average of the subject's interyear correlations for a time interval describes the process of *temporal stability* for that interval. This is the second of the two basic processes of identity formation previously described.

The results and analyses in the next section pertain to *groups* of subjects. The two racial groups are compared in terms of specific correlations as well as the average correlations. (A comparison might, for example, involve average values of many subjects and thereby include multiple single correlations.)

Thus, reviewing here the operational meaning of the identity measures and the general types of comparisons to be made serves to emphasize the complexity of these data. This complexity is inherent in the Q-sort instrument and, moreover, is desirable. As opposed to a single "objective" self-image study, this instrument is intended to analyze multiple self-images as they integrate at any one point in time and as they change over time. Identity formation is a complex phenomenon which refers to issues of integration and continuity, issues which are at varying levels of individual awareness.[4]

Presenting the sets of self-images in the first part of the chapter allows for observation of selected areas in identity formation. Processes of integration and continuity occur in each of these areas. When we then turn to the average correlations, it is the processes themselves that become foremost in our attention.

In both types of observation, the primary interest rests upon interracial comparisons. Do the black and white adolescents show different forms of idently formation? And, if there are differences, what do they consist of?

individualized decks for each subject, stems from Prelinger as described in NIMH Grant Proposal M-3642. His contribution to the basic methodology of this study obviously cannot be overestimated.

[4]Among the many arguments for the utility of the Q-sort is that through its many cards not merely the most prominent conscious tendencies are tapped. Other patterns, less prone to conscious control, also manifest themselves in the sorts.

DATA AND ANALYSES

Parental Images

One self-image set studied by the Q-sort was that concerning parental figures and the subject's relationship to them. Erikson and others cite the importance of parental identifications to identity formation. Aspects of these parental identifications were explored here through study of the subjects' idealized self-images with reference to these figures.

Each year the subjects were required to sort the features for themselves according to what they would be like if they were a perfect son to their mother *(ma)*.[5] Four sorts later, they were requested to sort the same features according to what they would be like if they were a perfect son to their father *(pa)*. The subject's first sort, it is remembered, is always with reference to what he is like right now *(me)*.

The intrayear correlations obtained refer to the relationships between the individual's present self-image and that which he believes would most suit each of his two parents. In addition to this, another relevant intrayear correlation concerns the coherence between the *ma* and *pa* images; in other words, how closely does that ideal self-image for father resemble that for mother?

Changes over time were also studied. There are, first of all, the variations over time in the intrayear correlations among these images. The other measure of temporal change is that of variations in the content of the images over differing time intervals, the interyear correlations.

Results

In comparing the two racial groups, the blacks showed consistently higher correlations than the whites between *me* and *pa* at significant levels (.05) in Year 1 and 2.[6] In later years the correlations for the two groups become more similar, but the values persist as higher for blacks. In figure 1, both higher values and less overall change for the blacks is apparent.

Going on to the relationship between *me* and *ma*, we again find higher correlations for the black adolescents. Here, however, the gap

[5]Symbols in parentheses following the descriptions of a given self-image refer to the notation used in the text, graphs, and tables.

[6]See Table 1 in Appendix A. The Mann-Whitney U Test was used to estimate this significance level.

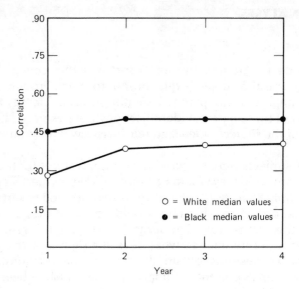

Figure 1. *me/pa* Intrayear correlations.

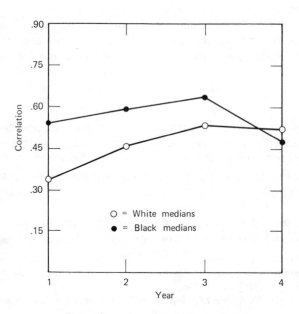

Figure 2. *me/ma* Intrayear correlations.

becomes smaller over time as seen in Figure 2. By Year 4 the black and white boys have almost identical median correlations. This is in contrast with the *pa* image, in which the blacks maintain a consistently higher correlation with their *me* image than do their white counterparts (Figure 1).

Still another view of the black high coherence and low change rate can be gained in comparing the groups for coherence between *ma* and *pa* (Figure 3). Here again the blacks show a higher and flatter curve than the whites, but in the fourth year the blacks show a sudden decline in the correlation between *ma* and *pa*.[7] It is the whites who develop a gradually increasing coherence between *ma* and *pa*, a steady increasing integration of the two self-images. The blacks, however, change less for the first three years and then show a sudden decline in coherence as they are about to leave high school.

In Figure 4, by way of triangular representation, the relationships for *ma*, *pa*, and *me* are depicted. The distance between two self-image points in a triangle is directly proportional to the size of the correlation between them. Most striking here is how the yearly triangles remain virtually congruent for the blacks. The white triangles, however, progressively increase in size over the first three years. The distinction

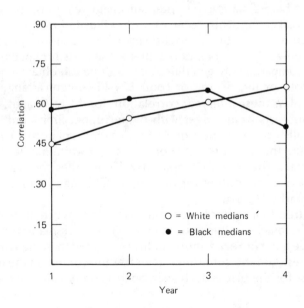

Figure 3. *ma/pa* Intrayear correlations.

[7]Use of the Mann-Whitney test also suggests this reversal; see Table 1 in Appendix A.

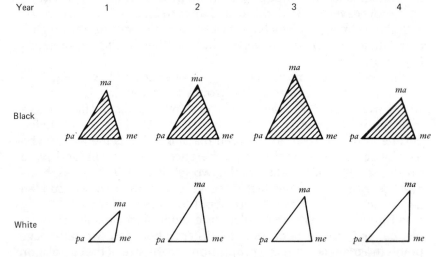

Figure 4. Parental images: intrayear correlations. *Note*: Length of line is proportional size of correlation between the two self-images involved.

between the blacks' stasis in parental self-image relationships and the whites' gradual change thus receives further illustration.

Additional evidence of the blacks' *lack of change* is in the content variation studies. It is remembered that if the sorts for specific self-images are compared between different years one can obtain a quantitative estimate of the degree of continuity in content arrangement. Figure 5 displays these interyear correlations for the *pa* images. For *pa* the black subjects have ranges with higher upper limits and higher medians in all but one comparison.[8] There is a similar pattern for *ma*, but for this image the disparity between black and white medians is not nearly as great as that for *pa*. Compared with the whites, the blacks have a higher stability of content for both parental images; this quality is especially marked for *pa*.

The dual trends of high coherence and lack of change recur again and again for the blacks in this parental data. The black *ma* and *pa* self-images are, except in Year 4, more highly correlated than the same pair for the whites. And the coherence as well as the content of the images changes less for the blacks each year. Such a combination of trends is

---

[8]As can be seen in Table 2 in Appendix A, these higher median values show up as positive tendencies in all intervals, and as highly significant for the Years 2-4 interval as determined by the Mann-Whitney U Test.

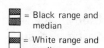

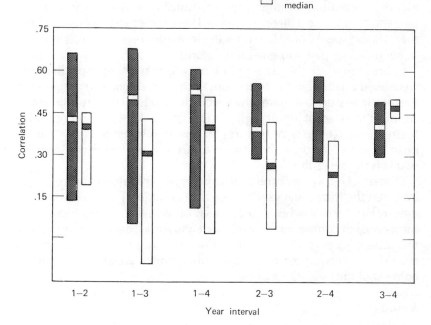

Figure 5. Range and medians of interyear correlations for *pa*.

reminiscent of the operational definition of identity foreclosure given earlier, that identity formation variant in which intrayear and interyear correlations show little or no change.

A further finding in these analyses concerns *differences* between the two parental images: (1) The correlation between *ma* and *me* is always higher than that between *pa* and *me* for the blacks, but it also fluctuates more than the corresponding *me/pa* correlation. Thus, by Year 4 the *me/ma* value is equal for blacks and whites (Figure 2); (2) In addition to the greater tendency for *ma* to change in its intrayear correlations, its content is also more subject to yearly fluctuation (see Table 2 in Appendix A); (3) Finally, the black *ma* image shows higher interyear correlations than does the white *ma*, but the discrepancy between the two racial groups is not nearly as great as that for *pa*. We return in Chapter 5 to consider the significance of these interesting differences between parental self-images.

Personal Time

A second self-image set studied was that of personal time. Primary identification figures, such as parents, represent one parameter of identity formation; another is the relationship of the individual to his past and future. Most directly involved here is the "sense of continuity" that Erikson speaks of.[9] How does the individual sense his relationship to his receding past and emerging future?

In order to study this aspect of identity formation, two specific sorts were used together with the *me* sort. Each year, subjects were required to sort the features of themselves according to how they recalled they were "at the start of junior high school" (*past*). A second sort required them to arrange the same features for how they imagined they "would be in 10 years" (*future*). The sort for *future* was the fourth one, that for *past* was the seventh.[10]

One set of intrayear correlations obtained here refers to the relations between the individual's self-image at present (*me*) and the two other time-related self-images. A second set of intrayear results focuses on the degree of coherence between *past* and *future* each year, reflecting the subject's personal time integration. A third group of correlations consists of interyear correlations, emphasizing degree of change in the content of the two self-images.

Results

Comparing the blacks and whites on the relationship between *me* and *future*, we find that the blacks show striking trends toward higher correlations in the middle years of the study. This is seen in Figure 6, where an initial similarity between blacks and whites gradually widens; by the fourth year the values have again become similar.[11] While in the process of going through high school, the black adolescents show a very strong resemblance between present and future self-images; but at either end of the process this coherence does not differ significantly from that of the whites.

In reviewing the relations between *me* and *past*, two very interesting trends appear (see Figure 7). First, there is a configuration of rapid change on the part of the blacks. Over the initial two years they show stable and high values; but then their correlations between *me* and *past* decline rapidly so that by the end of the study the median value is .10, an

[9]See, for example, E. Erikson, *Identity: Youth and Crisis*, 1968, p. 169.
[10]Since *past* refers to a fixed point in time, and *future* to a point always 10 years later, one would expect discrepancies between the two images to increase each year.
[11]Mann-Whitney tests confirm this finding; see Table 3 in Appendix A.

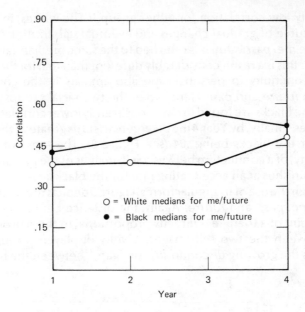

Figure 6. *me/future* **Intrayear correlations.**

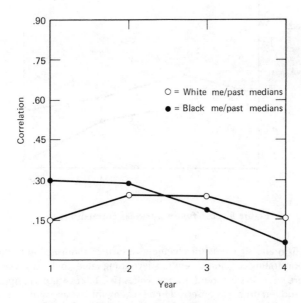

Figure 7. *me/past* **Intrayear correlations.**

extremely low correlation for either group.[12] The whites, in contrast, show a curve of gradual changes and eventual stabilization. By Year 4 the white *me/past* value has returned to the same median value as that in Year 1; this is a result considerably different than that for the blacks.[13]

A discontinuity in personal time also appears in the correlations between *future* and *past*. Here, when the two racial groups are compared, the blacks are found to have consistently lower correlations than the whites. Finally, by Year 4 the divergence is at its greatest; the median value for the blacks being .04 (see Figure 8). The sense of personal continuity, of a connection between self-images of the past and future, thus diminishes at an accelerating rate for the blacks.

Through the use of triangular forms, Figure 9 depicts the relationships among *me*, *past*, and *future*. Again, the distance between each self-image point of a triangle is directly proportional to the size of correlation between the two self-images. Vividly displayed through these figures is the growing *discontinuity*, or "gap," between the black *past*

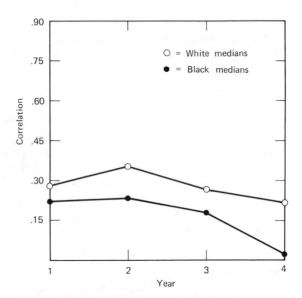

Figure 8. *past/future* **Intrayear correlations.**

[12]The Wilcoxin and sign tests for change give further confirmation to this graphic decline, demonstrating continued trends to decreasing values for blacks in later years in *me/past*. There were no such trends for the whites (see Tables 4 and 5 in Appendix A).

[13]A similar sudden drop was observed in discussing the *ma/pa* values for the blacks, raising the "brittleness" issue, of whether a foreclosed identity variant might be more likely to show tendencies to diffusion.

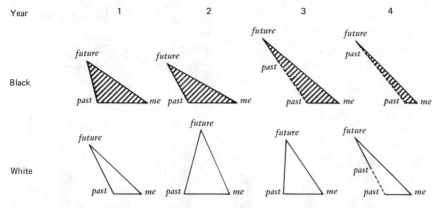

Figure 9. Personal time. *Note*: **Length of line is proportional to size of correlation between the two self-images.**

and both *future* as well as current self-image (*me*). Such a consistent and marked trend is not present for the whites.

The same theme of historical discontinuity is reflected in the inter-year changes in content (see Figure 10). For the *past*, the blacks show significantly greater discontinuity of content than do the whites. It is of interest that this difference varies directly with the number of years included in a particular interval, for example, in the interval between Years 1 and 4 the black median is considerably lower than in the interval between Years 1 and 3.[14]

For the *future* image a black discontinuity pattern is no longer present. Now interyear correlations are again higher for blacks and in fact they increase in direct proportion to the number of years in an interval. All black values are high; all medians are above .50. In contrast to the *past*, the blacks once more show a sameness of content, while the whites have greater lability of content in this image. The fluctuation in content varies with time for whites. The white subjects show many changes in the content of this *future* self-image, as opposed to increasing similarity for the blacks.

This difference between *past* and *future* can be further illustrated if one compares black *past* and *future* in terms of content change (inter-year correlations) as shown in Figure II. It is abundantly clear here that the *future* is a more stable self-image than the *past* in *all* intervals; such a difference is absent for the whites.

In reviewing this second self-image set, two important consistencies

[14]See also Table 6 in Appendix A for further documentation of this pattern.

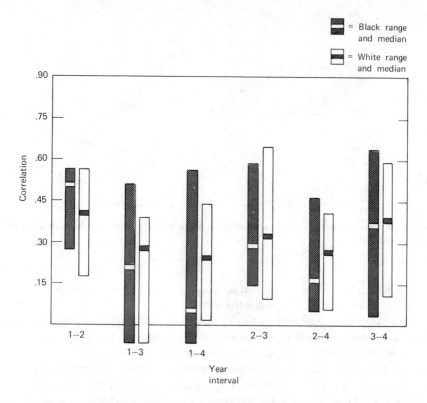

Figure 10. Ranges and medians of interyear correlations for *past*.

have been disclosed. The first is really further amplification of the foreclosure trends that emerged in examining the parental findings. For the black *future* self-image there is a pattern of sameness of content and high intrayear correlations.[15]

The second consistency appears to be the reverse of the first one. For the *past* the blacks repeatedly offer evidence of a break—a sudden decline in relevant correlations. This appears in three different contexts: their decrease in *me/past*; their decrease in *past/future*; and finally, their decreasing interyear correlations for the *past*. It is remembered that there was a suggestion of this phenomenon of sudden decline in the preceding section, as black *ma/pa* coherence decreased rapidly in Year 4.

[15]The averages for *future* reflect the same higher values for the blacks, see Table 3 in Appendix A.

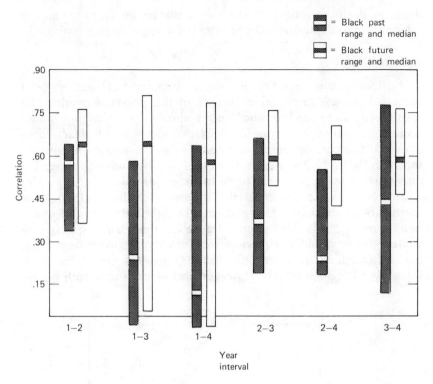

Figure 11. Ranges and medians of interyear correlations for black *past* and *future*.

At this point there are two different sets of self-images in which tendencies toward variant identity formations are *suggested*: the *foreclosure pattern* for parental images and *future* and the *diffusion pattern* for the *past* and for the *ma/pa* coherence. It is the black subjects who manifest these patterns in all the instances.

The Fantasy Self

A third set of self-images involve the individual's image of himself with a minimum of super ego restrictions, and emphasis on ego ideal. If there were no rules, inhibitions, or barriers, how would he imagine himself to be? What is the *most desirable*, most ideal image the individual holds of himself?

To explore this area the individual subject was asked to sort the features according to how he imagined he would be "if all your dreams were to come true; if everything would work out the way you want it to" (*fantasy*). Both the various vicissitudes of this image over time and its relationship to the individual's present self-image, *me*, were studied.

Results

The findings once again emphasize the unchanging characteristic of black self-images, a recurrent theme in the previous results. The relationship between *me* and *fantasy* shows a consistently higher correlation in all years but one for the blacks. Moreover, the black curve of *me/fantasy* correlations is considerably flatter than the white one as shown in Figure 12. The whites, however, vary in positive and negative directions showing an overall trend to increasing value.

The content variations are similar in form, with the black interyear correlations for *fantasy* being consistently higher than are the whites (see Figure 13).[16] Also illustrated in Figure 13 is the related aspect of the whites' greater variation in medians from one interval to the next. Thus, the white medians vary from .56 to .35 and generally seem to diminish in value over longer intervals. However, the blacks show *both* higher

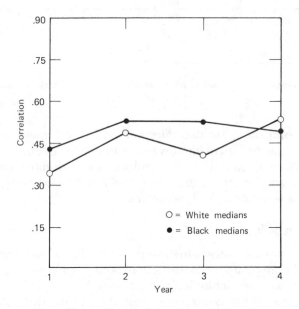

Figure 12. *me/fantasy* **Intrayear correlations.**

[16]This is confirmed by the Mann-Whitney test as all but one of the black values show tendencies in this direction and significant results in the Years 2-4 interval as seen in Table 7 in Appendix A.

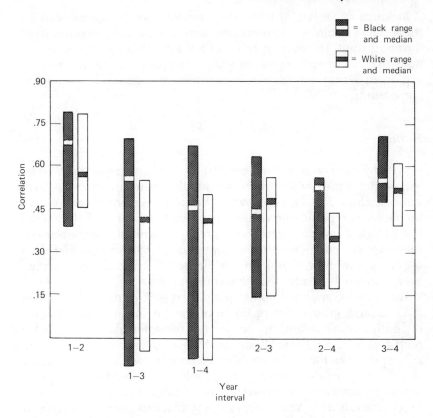

Figure 13. Ranges and medians of interyear correlations for *fantasy*.

medians and less fluctuation in these medians over the entire four-year period.

The black trend toward identity foreclosure receives further emphasis in this third set of self-images. For the black boys the *fantasy* self-image is relatively nonmalleable in content. It is also more tightly connected to the self-image *me*.

The Present Self-Image

The present self-image has been referred to in all of the preceding analyses. The relationship between present self-image, *me*, and the other images served as a standard by which pattens of intrayear correlations at single points as well as over time could be observed and compared. However, it is also relevant to discuss the comparative studies of this self-image.

In terms of structural integration aspects, the image *me* can be studied by examining its annual averages of intrayear correlations for the two groups. This average represents the mean of the seven correlations that occur with *me* each year.[17] Temporal variation for *me* is expressed by the time-related changes in the averages and the interyear correlations.

Results

In many respects the image *me* differs from each self-image thus far explored. For the first time there is a virtually identical configuration of values for *both* racial groups. When the medians of intrayear averages for *me* are compared, the values are strikingly similar, with the exception of Year 2 when a suggestion of greater magnitude for the blacks is present.[18] The curves for *me* are more parallel than for any other set of self-images we have examined (see Figure 14). The black curve, however, is somewhat flatter than the white one.[19]

The content variation findings also display striking parallels between the two racial groups (Figure 15). Here, the *me* image shows a gradual discontinuity in content between the first two interval comparisons for both groups as manifested by diminishing interyear correlations. Over all later intervals the *me* image shows increasing continuity of content, a reversal of correlations which is true for *both* blacks and whites.

Also present here is the black tendency toward sameness, however. In all intervals from Year 2 onward, the black medians are markedly higher than those of the whites. Thus *me* becomes more constant in content for both races, and for the blacks it remains at even greater levels of constancy over these three later intervals.[20] The whites, however, show a gradual steplike rise in these interyear correlations until by

---

[17]That is, *me/ma, me/friend, me/future, me/pa, me/other, me/past, me/dream*. To obtain this mean, each of the correlation coefficients was first transformed to a z value (Fisher transformation) and these converted values were then averaged. All average correlations referred to hereafter in the text were calculated in this way.

[18]This deviation from the pattern here supports the previous noted tendencies of the black values. Further statistical confirmation of these results is shown in Table 8 in Appendix A. It should also be noted that if the intrayear averages of other self-images are examined the parallels found here are not present, hence the similarities are not simply a consequence of comparing averages. The groups differ in the same manner as in the specific intrayear correlations observed in the previous sections.

[19]Both the suggestion of greater magnitude and the flatness are consistent with the recurrent findings for the black subjects: intrayear correlations and low change rate.

[20]See also Table 9 in Appendix A for further statistical confirmation.

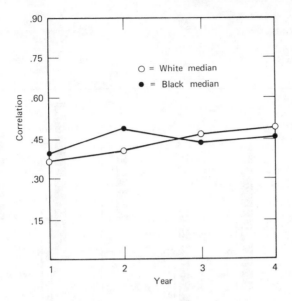

Figure 14. Average *me* intrayear correlations.

the Years 3-4 interval their median approaches a level similar to that of the blacks.

It is for the self-image *me* that the two groups are least differentiated from one another. However, they are *not* identical. There are subtle indications of differences between blacks and whites, again in the direction of foreclosure patterns for the black boys.

Structural Integration and Temporal Stability

In order to measure the overall degree of self-image integration for each subject, the average of all his intrayear correlations was calculated each year. To determine the degree of self-image content stability over a given time interval for a subject we calculated the average of all his interyear correlations for that period (e.g. Years 1-3). In terms of our identity definition, these two averages are, respectively, the primary indices of *structural integration* and *temporal stability*. The preceding sections reviewed various facets of identity formation as revealed through specific sets of self-images. In now shifting to analysis of these averages, we directly confront the overall processes of identity formation. In other words, we now return to the basic question, do the two racial groups differ in their identity formation?

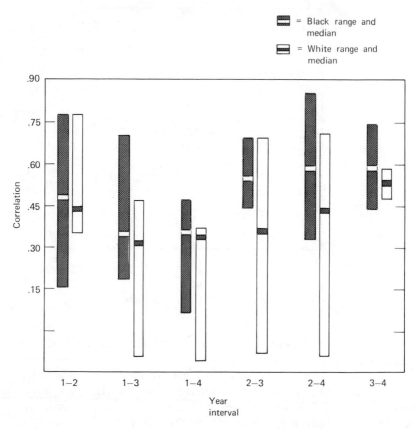

Figure 15. Ranges and medians of interyear correlations for *me*.

Results

The average of each subject's **correlations** for any given year was calculated from his 28 separate intrayear **correlations** for that year. The absolute value of this average represents the fit of all self-images for that year. A value of 1.0 means that all the subject's **self-images** were exactly the same in that year, while a result of 0 **represents** absolutely no resemblance of the images to one another in **that year.**[21]

[21]One further possibility exists, and that is a negative average **correlation, which** would represent a reversal of content. Such an average never emerged, **although** there were individual correlations of this nature, as is discussed later. Related to **this is the fact that an** average of 0 could represent a combination of negative and positive **correlations. This**

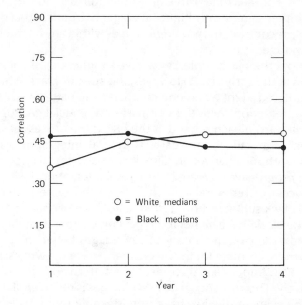

Figure 16. Average intrayear correlations: structural integration.

It is this average intrayear correlation that operationally defines a basic process of identity formation, namely, structural integration. Change in the value of the average over time is a reflection of changes in the direction and rate of structural integration. The two racial groups clearly differ from one another in terms of this process. The black intrayear averages show remarkably little change from year to year. Figure 16 reveals the rather flat curve for the black. The Wilcoxin test for change shows *no* significant black changes in this average over any time interval.[22]

The whites, however, display a steadily rising structural integration curve. Moreover, use of the sign test on their values suggests change over four different intervals.[23] Still further evidence of these racial differences in structural integration is given by application of the Mann-Whitney test, which compares yearly sets of average intrayear

would also mean that the *overall* fit of images was at a 0-level despite the wide variability in specific self-image partitions. See footnote 17 for comment on the calculation of these averages.

[22]The sign test, however, does suggest increasing values in two intervals; see Table 10 in Appendix A.

[23]See Table 10 in Appendix A.

correlations. For the whites three different intervals show the later year as having either significantly higher averages or trends in that direction. The blacks, in contrast, show *no differences* within any interval in terms of these averages.[24]

In addition, the black adolescents have a higher degree of structural integration in the early years. However, this specific black-white difference is rendered slight when one compares the patterns of *change* in structural integration. With this comparison the racial groups become differentiated from each other. For the blacks the process of structural integration is a static one; yearly structural integration values are essentially without change. In direct contrast, the whites display a process of *progressive increase* in structural integration.[25]

The process of temporal stability reflects this same pattern. Once again, the black subjects show a striking *absence* of change. In all time intervals the blacks have higher interyear correlations than do whites. In other words, the overall *content* of self-images between years is emphatically more constant for the blacks than it is for the whites. Figure 17 and Table 12 in Appendix A document this pattern both graphically and quantitatively. The content of their self-images is relatively fixed for the blacks. However, the whites have unmistakably more malleable self-images. This greater flexibility in white self-image content is reflected in their lower interyear averages and in their generally broader range of average correlations in each interval.

To summarize, it is the blacks' relentless sameness—the "flat line" quality—that is forcefully described by these important averages. Their structural integration, a process earlier defined as the average intrayear correlation, shows minimal variation from year to year. In addition, at several points in time, the actual degree of integration is higher for the blacks.

The other basic process of identity formation is temporal stability. This is the variable defined by the average of interyear correlations. Its value was *always* higher for the blacks. Over all time intervals, then, the black subjects have greater constancy of self-image content.

The values for structural integration and temporal stability are both lower and show a greater tendency to fluctuate for the whites. In terms of identity formation variants, the blacks clearly exemplify identity foreclosure. This variant was earlier operationally defined as being present when "either the structural integration and temporal stability of self-images remain stable; or only temporal stability shows con-

---

[24]See Tables 10 and 11 in Appendix A.

[25]See Figure 16; Table 11 in Appendix A gives statistical indications of this difference as well.

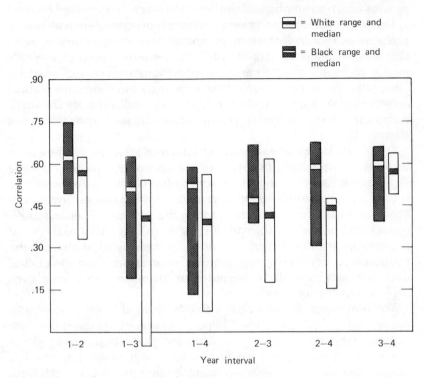

Figure 17. Ranges and medians of average interyear correlations: temporal stability.

tinued increase, while structural integration is unchanging."[26]

Of the two identity formation processes, it is that of temporal stability in which the black adolescents differ most from the white adolescents. The black subjects show greater constancy of self-image content over time during every possible year interval. Additional suggestion of the blacks' static content was given in examining the results for the separate sets of self-images. For each image except *past*, the black interyear correlations were higher than the whites. This difference is even found for the *me* image as well, a self-image in which the racial groups differed less than in any other. The overall black pattern, when both identity formation processes are considered, is consistent with the operational definition of identity foreclosure. However, it remains an unanswered question as to what the significance is of temporal stability being that

[26]The fact that this variant, as well as the other identity variants, is defined in terms of comparisons over time is discussed in Chapter 2, especially p. 36.

process which so emphatically reflects the blacks' static, fixed quality.

The whites exemplify an overall pattern most consistent with that of progressive identity formation, in which "both the structural integration and the temporal stability of . . . self-images are simultaneously increasing over any given period of time, though not necessarily at the same rate."[27] It is remembered that from Years 2 to 4 there is a gradual increase in the magnitude of white temporal stability (Figure 17). There is also a steady increase in white structural integration from Years 1 to 4 (Figure 16).

These trends are in the same direction as those discussed in the earlier sections, in which specific sets of self-images were explored. In those analyses we found strong suggestion of foreclosure patterns for the blacks, and progressive identity formation for the whites. The significance and basis for these black and white patterns in separate self-images and in identity formation remains to be considered. Also of importance are the clinical questions. Namely, what, if any, are the relationships between these quantitative measures and the clinical data, the longitudinally collected case history material from each subject over the four years?

The remaining chapters deal with the issues of interpretation, etiology, and clinical relevance. Chapter 5 examines the questions with respect to the sets of self-images. In the final chapters, possible roots of black identity foreclosure are discussed, relevant explanatory models are constructed, and recent empirical and theoretical work is reviewed.

---

[27]See footnote 26 regarding the nature of this definition.

CHAPTER 5

Interpretations and Clinical Parallels

INTRODUCTION

The next three chapters are closely related in aim. Together they take up the diverse findings presented in Chapter 4. Detailed analysis and interpretation of these complex results is approached in these chapters, beginning first with the data at hand, then concluding with theoretical models and implications.

In this chapter aspects of specific self-images are discussed in terms of clinical parallels as well as theoretical speculations. A similar approach is then applied to reviewing the identity formation processes themselves. Finally, we look at the question of how individual historical events are related to the quantitative findings. In response to this important issue of "developmental relevance," the concluding sections of the chapter follow the identity formation of a black and white adolescent through both historical and Q-sort data.

PARENTS AND IDEALIZED FIGURES

For the black subjects an important result appeared in the analysis of parental self-images. With respect to both parental self-images, the blacks show relatively unchanging and high correlations with the other self-images. In addition, both images have a relatively fixed content over time. Such trends emphasize the significance of parental figures in *both* the structural integration and temporal stability of black identity formation. There are also some interesting differences between the two self-images. The subject's ideal self for father is a most inflexible image: its correlations show but meager change in terms of both intrayear and interyear values. The ideal self for mother has generally high correlations with other self-images; but it also shows evidence of greater flexibility with respect both to relations with other self-images and in shifts of its own content. Because of this greater tendency to change on the part of the *ma* image, the initial rather high coherence between the two parental images undergoes a decline in the last years of the study for the blacks.

Seen alongside of our family data, the results for the black *pa* self-image are at first puzzling. Almost two-fifths of the blacks had no father at home;[1] most of the others spoke of their father as a disparaged, uninspiring man. However, on the Q-sorts the blacks have extremely high and unchanging correlations for the *pa* image. These correlations—both intrayear and interyear—are among the most rigid of the black values. One solution to this puzzle may lie in the following speculation: Given an absent and/or degraded father, his place in the identity formation of his son is that of a rigid unchanging figure whose expectations are imagined by the son always to be the same.[2] In such a situation, unavailable to the son is the chance to anticipate, compare, and modify the ideals he *imagines* are held by his father with his father's *actual* ideals. Through an immense lack of information, the paternal self-image becomes a foreclosed one, a "pre-packaged" one, and thereby a major contributor to the identity foreclosure of the son. This kind of sequence is discussed further in Chapter 6.

The interplay between parental images and other self-images for a subject was often instructive. At times, a shift in the intrayear correlations for *ma* and *pa* was paralleled by an increase or decrease in the corresponding values for the subject's *me* image. On other occasions the interaction was with parental images, future, and overall structural integration. Consider the two black subjects, Lenny and Jerome. For these adolescents the correlation between *me* and the two parental self-images significantly reversed their values over a one-year period. The *me/ma* value now became significantly higher than *me/pa* for both boys.[3] This change paralleled decreases in the magnitude of the *me/future* correlation and in the structural integration correlation. Clear historical events paralleled these quantitive shifts:

Lenny had been "persuaded" to go to college by his mother and guidance counselor. To be a Marine as was his father was suddenly becoming an unacceptable ambition in the eyes of his elders, and in part for him.

Jerome was with his mother's urging becoming increasingly committed to a college education and the career of a physician. Continuous sources of anxiety lay in the facts of his dubious academic performance and in his father "staying out" of his decision. Father was a postal clerk. Mother, about to receive her night school degree in education, clearly knew better about these matters.

---

[1] In contrast to the whites, one-tenth of whom had no father at home.

[2] This speculation is supported by the fact that when blacks with fathers were compared with blacks without fathers the latter group demonstrated consistently *higher* intrayear and interyear correlations for the *pa* self-image. See Tables 13 and 14 in Appendix A for a summary of the Mann-Whitney tests on these values.

[3] At $<.05$ probability level as determined by the *t* test.

Table 1. Shifting Intrayear Parental Correlations for Two Black Adolescents

Subject and Historical Event	Year	Intrayear Correlations		Structural Integration
		me/ma	me/pa	
Lenny	1	.39	.63	.39
Increasing college	2	.69	.61	.44
pressure; anti-	3	.65	.44	.31
Marine ambitions	4	.68	.59	.36
from mother and				
school				
Jerome	1	.69	.70	.54
Increased medical	2	.53	.46	.48
school wishes				
coupled with poor				
academic perfor-				
mance and greater				
urging from				
mother to medical				
school ambitions				

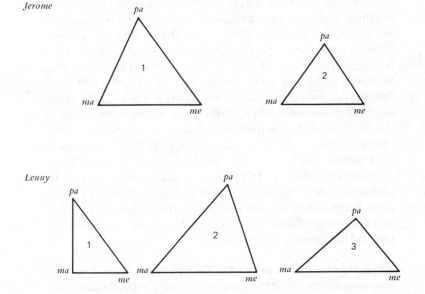

Figure 1. Intrayear parental self-images: two black adolescents. The distance between points of a triangle is directly proportional to the correlation between the two self-images. The number within a triangle indicates the year of the Q-sort.

CONTINUITIES AND DISCONTINUITIES IN PERSONAL TIME

Relationships between personal time and identity formation were studied by means of the self-images for *future* and *past*. For the *future*, high intrayear and interyear correlations were repeatedly found in the black subjects. This self-image always differed from that of the whites, and in the same direction, that of *greater rigidity* in terms of both its relationship to other self-images and its content repetition.

However, the *past* self-image disclosed a new pattern for the blacks. For these subjects there was now a persistently low interyear correlation, thus indicating a marked *discontinuity* of content. In addition, the blacks always showed lower correlations of the *past* with the current self, *me*. Extending the impression of diminished intrayear correlations were the values for black *future* and *past* coherence. These values were also strikingly lower than those of the whites. This set of trends is remarkable in that it represents a *total departure* from all other findings for the blacks.

Some clues as to the significance of this deviation can be found in the interview material:

One of the most dramatic events in Frankie's adolescence was his trip to the South. It was through this visit that he made the startling and unhappy discovery of how unacceptable his ancestors and current relatives were. The disappointment and chagrin were aggravated by the multiple rejections he then received from these "farmers," who disliked and distrusted him because he was from "up the road" (the North). Although Frankie presented the most obvious example, a similar problem was observed in each of the black subjects. The sense of discontinuity with the past of their family and, more broadly, race was present at all times in varying degrees. The predominant pattern was the subject's ready denial of any knowledge of his family's history. The subject would claim to know little about his father's background, regardless of whether or not he were living at home. He knew much more about his mother and her present family, but had little to say about either family beyond his parents' generation.

A second expression of this theme was in the blacks' derogatory comments about the "southerners," the blacks who were "stupid" and "out of it." Half of the black subjects were born in the South, and all had relatives still living there.

There was much to indicate shame of all aspects of their past. Benny, at once annoyed and embarrassed, directly criticized his ancestors for being "slaves" and "putting up with it." Another expression of the blacks' rejection of their past was in their violent dislike of the Black Muslims. Besides urging segregation, the Muslims emphasized the issues of racial history. The subjects viewed the movement with distaste: they were troubled by the "stories" the Muslims told which had no connection to "reality."

The white subjects gave many indications of their experienced continuity with their past:

All of the whites knew the history of at least one previous generation on both their mother and father's sides. Some knew of several sets of grandparents, of their life and work in Europe and the United States. There was neither reluctance nor uneasiness in speaking about these events and figures of the past. Indeed, several white subjects were extremely proud of their relatives, admiring their possessions and reputations. The disagreeable elements that must have existed were greatly overshadowed by the subjects' proud memories and abundant factual knowledge of the past. Most importantly, the negative aspects of their histories were not of sufficient weight to even suggest any disruption or denial of a past.

A number of determinants underlie the discontinuity expressed here by the blacks in both the quantitative and qualitative data. One, already alluded to in many ways, is their shame of the past. A second is the actual instability in the black families. Many of the black subjects had minimal acquaintance with their father and his family. Although these same subjects appear to have greater overall knowledge of their mothers, it is questionable as to how much contact these women maintain with members of the preceding generations. The white subjects emerged from relatively stable paternal and maternal lineages, changed residence much less frequently than the blacks, and were often physically surrounded by relatives of several generations. Some spent most of their earlier years in the context of a large extended family within the same house.

Associated with the phenomenon of black historical discontinuity is the issue of black self-hatred. This issue represents a major source of the black subject's shame of his past. It includes the history of slavery, exploitation, rejection, and paternal desertion. It includes the personal history of being "blood" relatives to a series of undistinguished often degraded men. Thus not only were our subjects members of a generally ostracized and disrespected minority, but they were also among the "undesirables" within that group: *men.* The black boys at times spoke intensely of this self-hatred. Although dark-skinned and of southern vintage, they insisted upon their distaste for dark-colored blacks and for "stupid" southern blacks.

In one boy, Benny, the expression was even more direct. To Benny, most blacks were "coons," "cats who have been 'bad' and drank too much." The blacks confirmed for him "the image" that society had of him and his people: lazy, drunk, "no-good men."

These observations of the rejection of the past and the associated self-hate coincide in part with the findings of Kardiner and Ovessey[4] in their emphasis on self-hate. However, we found no evidence to support their conclusion regrding the blacks' desire for "whiteness," a wish Kardiner and Ovessey suggest as strongly associated with the self-hate. Such a wish was not discernible in the extended interviews, projective materials, or Q-sort data. In all of this data there were indications of varying degrees of self-rejection coupled with self-resignation.[5] Wishes for who they wanted to be are most closely related to the dreams and wishes of the subjects, to their inner heroes. This area was studied through the Q-sort by use of the image for fantasy. Although the results do not confront the issue of "whiteness," they do speak directly to the question of resignation.

FANTASY AND FAILURE

Analysis of the black patterns for *fantasy* image disclosed trends already noted for parental as well as *future* images Intrayear correlations of *fantasy* image with current self, *me,* are consistently high and unchanging. The measures of content change, the interyear correlations, indicate that the *fantasy* image itself is considerably less malleable for black boys than it is for whites.

Again, it is valuable to turn to the interview data for some clues to the understanding of why the blacks display here a *fixed* image both in terms of structural relationships and content:

Frankie, a black subject, had few positive conscious or pre-conscious identifications. There were fleeting glimpses in the interviews of his brother-in-law, uncle, and southern grand-

---

[4] A. Kardiner and L. Ovessey, *Mark of Oppression,* 1951.
[5] The gamut of implications for self-degradation and self-hate are discussed more fully from the standpoint of an explanatory model in Chapters 6 and 7. We consider new empirical data and conceptualization about black self-esteem in Chapter 8 and 9.

father, and even these men were said to have many faults. Most of the blacks frankly stated that they had no heroes, no people they wanted to resemble now or at any other time. They wished "just to be myself." Occasionally, uncles, teachers, prominent black athletes, and black small businessmen were selected as ideal figures. Black responses were not so sparse when the question turned to antiheroes, those people whom they wished above all not to resemble. Readily listed were bums, beggars, gangsters, and thieves. In some cases, such as Frankie's, the most prominent antimodel was his father.

There were interesting variations on this hero-antihero pattern:

Lenny frequently wished to "take after" his father, to be a Marine. However, the community and his mother seemed to intervene as they urged him to richer "opportunity" and "challenge." Thus one of the few blacks who found a positive hero in his life, his father, was forced to reject this man in favor of new, and to him unknown, images of black successes.

Jerome wanted for many years to be a "doctor." Despite an unremarkable academic record and difficult school adjustment problems, this wish remained strong. His father, a postal clerk, was rarely mentioned in any of the interviews. When Jerome did speak of him it was always in an uneasy embarrassed manner.

Benny was a third variant. He openly and repeatedly berated his "mean" and "stubborn" father. This man, separated from the family for many years was seen by Benny as an outcast, an undesirable. Benny insisted that the only reasons for any visits to the man were for "cash."

Given what appears to be so great an impoverishment of positive models, it does not seem surprising that the black self-image of what he would be like if all dreams were realized was a remarkably *static* one. It was static in relation to both its content over time and its relation to the current self-image. There was but little choice available. Yet the ingredients for fantasies are plentiful. It may well be that the ease with which dreams can become nightmares for these subjects was sufficient to greatly restrict their flexibility in considering *any fantasy* self-image.

There is still a third possible reason for the restricted nature of the black subjects' *fantasy* image. Once more, it is suggested in the interviews:

The status of mothers for the blacks was different. Unanimously, mothers were described with the strongest superlatives. They were advisers, exceedingly nurturant providers, and "bosses." Any problems, from the most trivial to the most profound, could be solved by them. They might also initiate major conflicts for their sons in their authority roles. Subjects often worried about their "moms" and their recent report card. To speak of heroes for the black boys may, then, be inaccurate. A major figure of conscious admiration and virtual worship was a heroine mother.

One can only speculate as to how threatening such a "hero" may be for the black boys. One way to deal with the threat of a woman as an ideal image is to tenaciously hold to a consciously acceptable ideal fantasied self-image (of a man) and to permit absolutely no variation in it.[6]

The white subjects revealed much flexibility in intrayear relationships and malleability in content for *fantasy*. Accompanying this opposite quantitative pattern is a very different clinical picture as well:

There were an overwhelming number of heroes for the whites. The men they admired varied from history teacher to astronaut. Not only was there an abundance of such men, but the number continuously grew. Almost yearly, these boys discovered new people they wished to emulate either by direct acquaintance or via the mass media. The blacks, in contrast, only occasionally found a new model. Most often their new ideal figure was a neighborhood bachelor or football star. And within a brief span of time this added hero would be "dropped" from the already small black repertoire.

Another feature of the whites' heroes was the seeming influence that they had. Two white subjects' changes in career plans occurred during and immediately after a year with new teachers of whom they were exceedingly fond. Most convincing, in terms of influence, was the fact that each of their new

---

[6]Here then is the notion that rigidity in a self-image may offer a form of insulation, protection, from conflicts inherent in the image. This is similar to the hypothesis that identity foreclosure serves such a such a function for the subject in general as discussed in Chapter 3. Other speculations about functions of identity foreclosure are discussed in Chapter 6 and 7.

goals was to become exactly what this new hero was. A third subject became increasingly motivated to attend college just after several athletic heroes decided to continue their education rather than enter professional athletics.

In addition to greater numbers of positive models for the whites, the nature of their antiheroes also differed:

> There were usually specific people whom the white boys wished to avoid resembling. Other than the generally disliked "garbage man," all other figures were associated more with the individual subjects and his tastes than with being white. Joey, for instance, was unique in his hate of the "Boston Strangler" and Lee Oswald. Others chose thieves, murderers, cheats, bigots, and "idle men." There were few antiheroes for each subject. In fact, some subjects claimed to be unaware of anyone they did not wish to resemble. In addition to being small, the individual list generally remained stable over the years. Detested figures were rarely added or deleted.

Using the dual assets of greater availability of positive models and fewer negative models, antiheroes, the whites maintained a *fantasy* self-image capable of numerous changes. Absent for them was both the impoverishment of models and the need for insulation through rigidity.

THE CURRENT SELF-IMAGE: ME

Qualities of rigidity and unchanging content are no longer present to any noticeable extent for the current self-image, *me*. Curves of *intrayear* correlations for *me* over the four years are virtually identical for both racial groups. Not only are the changes in intrayear correlations very similar for blacks and whites, but the magnitudes of the correlations are also near equivalent. Moreover, the content variation findings show many parallels; *me* becomes more constant in content for *both* races.

These findings are surprising and extremely interesting. For black and white subjects, the current self-image undergoes very similar vicissitudes in its correlations with the other self-images and in its own content. There are but undertones of the black *me* having greater constancy of content and stronger similarities to their other self-images. Why then does this self-image show so little differentiation between the two racial groups?

In relation to parental ideals, personal history, and fantasy, the black adolescent showed trends suggestive of identity foreclosure. For his present self-image this configuration was only subtly hinted at.

There are several ways to consider this departure. A number of highly problematic issues are involved in the self-image sets already considered. Among such issues are disturbed parental identifications, with the black usually having an absent or degraded father; an uneasy and largely unwanted relationship of the black to his past; a restricted and conflictful perception of his future; and a fantasy life again influenced by restricted alternatives and the multiple anxieties attendant upon these restrictions. When not brought into direct focus, these disturbing issues are not highlighted in the data. Yet they all obviously impinge upon the subject's present image of himself, and this impingement, this influence, is revealed in many subtle ways as the *me* image is carefully examined. However, when the subject's attention is forced upon the problem areas, their importance and consequences become most clear, as the preceding repeating patterns for parents, personal time, and fantasy have so well demonstrated.

One may, with good reason, question why the impact of these issues upon the black boys is not reflected more clearly by the current self-image. For it is in this image that the subjects appear most similar to one another. The present self-image is the one of which the individual is most conscious and is also the most amenable to the influences of everyday events and stresses in his life. What he *consciously* "thinks of himself" may indeed be quite similar from moment to moment for the black and white youth. Both sets of boys attended high school, had varying groups of friends, went to parties, and were not in any obvious way depressed or deflated. When a highly self-conscious self-image is requested, "What you are like now," the possibility of less conscious conflicts being stimulated and tapped by the Q-sort is minimized.

The other self-images are likely to be less familar to the subject. The images are constructs he has rarely, or perhaps never, consciously entertained. Being less familiar, they are also less subject to conventional responses, those responses that would be more routinely given many adolescents. Without this shield of conventionality, the various self-image differences between racial groups become more readily apparent. In brief, requesting the unfamiliar runs counter to the option of a stereotyped response, a response cleansed of idiosyncratic shading or conflict. In situations of extreme pathology, marked confusion or self-hatred, for instance, we might expect that even the current self-image would reflect rigidities or confusion. By design, neither of our groups of subjects included such overt pathology.

Given these assumptions as to what the *me* self-image represents, two important methodological facts become very clear. First, differences in the two racial populations are not to be found in examination of *only* their current self-image.[7] Second, identity formation cannot be studied by means of a single self-image, *me*.[8] This position was insisted upon in the original definition of identity formation. The concept refers to specific ego development processes, both conscious and unconscious, involving a wide array of intrapsychic and psychosocial issues, and these issues range from those concerning libidinal needs to significant defenses to community recognition. Such an array of variables is surely not encompassed by the current, highly conscious self-image. At times Erikson discusses the relationship of concepts such as self and self-image to identity.[9] The position taken here is that self-image, as the term is ordinarily used—namely to refer to the current highly conscious image of one's self—is *not* a sufficient index of identity formation. However, it is also not irrelevant; this image must be studied as *one aspect* of any given identity formation.

In the analysis here the *me* self-image has been used as a form of "base line," an image to which all other self-images have been compared in viewing their intrayear correlations. It is in relationships, the correlations of *me* with these other self-images that less conscious, more conflicted issues have become apparent. Through such data, utilizing the other self-images, differences between the racial groups have emerged with increasing clarity.

THE PROCESSES OF IDENTITY FORMATION

No one set of self-images represents a person's identity formation although the different images reflect varying facets. It is the overall patterning of self-images that defines the person's identity formation. The two basic processes of identity formation are structural integration and temporal stability, operationally defined by two specific averages: those of the intrayear and interyear correlations.

The black and white groups diverged in respect to their interyear and intrayear correlation averages. The black intrayear averages were

[7]This fact is of course confirmed here. It also helps to explain why Rosenberg did *not* find any differences between racial groups in their "self-images" (personal communication); see M. Rosenberg, The Adolescent Self-Image, 1965. The point is, of course, pertinent to the many self-image studies which are reviewed in Chapter 8.

[8]Erikson makes this point: "(social scientists) . . . sometimes attempt to achieve greater specificity by making such terms as 'identity crisis,' 'self-identity' . . . fit whatever more measurable item they are investigating at a given time . . . they try to treat these terms as matters of social roles, personal traits, or conscious self-images, shunning the less manageable more sinister—which often also means the more vital—implications of the concept," Youth: Identity and Crisis, 1968, p. 16.

[9]See footnote 8 and Erikson (1959).

highly similar from year to year; there are no significant changes in this value over any time interval for them. In contrast, the whites' intrayear averages rise steadily. Over several intervals there is evidence of highly significant changes. In addition to their lower rate of change, the black averages were of greater magnitude. In sum, the findings are consistent with a static process of structural integration for the blacks. Taken at any separate point in time, the blacks show a higher degree of structural integration than the whites.

Interyear averages, indices of temporal stability, repeat this pattern. The blacks had higher interyear averages in all time intervals. Since interyear correlations measure shifts of self-image content, this finding means that the black subjects differ from the whites by their maintaining a greater constancy of content for their overall set of images. Once again, it is the blacks who show an *absence of change*, a stasis. Their self-images are distinguished by the tendency to repeat the same patterns of characteristics from year to year, thus resulting in high interyear correlations. The whites tend to vary in the yearly composition of their self-images, their interyear averages always being lower than those of the blacks.

The blacks, then, are distinct from the whites in terms of both basic identity formation processes. The black temporal stability is consistently high over the four-year period, and structural integration, the second basic process, fluctuates significantly less for them. This patterning of the basic processes conforms precisely to that of *identity foreclosure*. This identity variant is at the opposity pole from a moratorium, in which the individual appears in flux, "finding himself," and "getting his feet on the ground." Rather than any form of fluctuation, the blacks showed signs of having a clearly defined array of self-images. From at least the beginning of high school, these images have been unchanging in terms of both their interrelationships and their specific contents.

In many ways this static quality was articulated in the clinical data. It was very striking when black boys spoke of their "future":

> The black adolescents had little to say about the future. Generally, their alternatives were to work in a factory or perhaps join the armed forces. Or, if their dreams were fulfilled, they might some day own a small store and finally be "boss." Frankie, for instance, saw the future only unenthusiastically. He might have a job. However, just as likely were the possibilities of unemployment and a bad marriage. The sensed absence of choice, of "potentials" and "challenges," was more than obvious.[10]

[10]In the next chapter several other clinical illustrations of "stasis" qualities are given.

The whites, in having a lower degree of temporal stability and a fluctuating structural integration, suggested at several points the possibility of a psychosocial moratorium. Moreover, one white subject showed unequivocal evidence of such a state in his marked decline in temporal stability accompanied by a pattern of stable structural integration. Another white adolescent strongly hinted at a moratorium as his structural integration alternately rose and declined over the years. The general white pattern was most consistent with that of *progressive identity formation*, in which both the temporal stability and structural integration progressively increase in magnitude over time.

THE IDENTITY FORMATION OF TWO ADOLESCENTS

Both black and white subjects were presented with several unanticipated changes in their plans for future work and education. They were not oblivious to the surprises, nor were they able to fully insulate themselves from their impacts. Yet responses were distinctive, and illustrate in greater detail some differences in the identity formation patterns of the black and white adolescents. Two boys, already briefly referred to, provide excellent examples of a psychosocial moratorium and an attenuated diffusion. Here we look more closely at their development and its association with Q-sort findings.

> *Jim, a white, Italian boy.* Jim decided at the close of a year in trade school that this was neither the kind of training nor future work that he wanted. However, having begun the school, and with some parental pressure, he unhappily remained there. He continued to "try it," hoping that welding would become "better." By the end of the second year there he was surer than ever that he wanted to go to college and eventually become a history teacher, most certainly not become a printer. To the dismay and chagrin of his family, he began investigations into preparatory schools and colleges. The inner confusion and changes of this period are well represented on the Q-sort (Figure 2). Most dramatically, *all* self-images changed strikingly in content between the second and third years of high school. The greatest change in content was between the first and third years. At times the change was a complete reversal of priorities, as expressed by negative interyear correlations. Two sets of his interyear values are seen in Figure 2.
> The negative correlations, indicating the reversed self-image contents, persisted for certain images until the last two years. By the third and fourth years of the study, Jim had become most settled with his decision for prep school and a

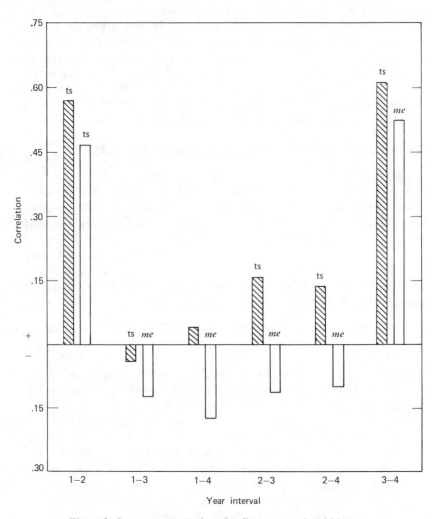

Figure 2. Interyear correlations for Jim: temporal stability.

career of history teacher. This newly found certainty is reflected in the set of correlations for the interval Years 3 to 4. In this interval all correlations now return to highly positive values.[11]

---

[11]Had the interyear correlations—particularly their average—remained negative until the end of the four years, Jim would have exemplified our single clear case of negative

Jim remained overtly "stable" despite the turmoil in his plans and the major shifts in the content of his self-images. This clinical presentation was also expressed in his Q-sort results (Figure 3). Jim's average intrayear correlations for *future* and *me* declined slightly in the first year and then stabilized or increased in the last two years of the study. His average intrayear correlations, the index of his structural integration, showed the same cycle of variation, as seen in Figure 3. Thus, there was a decrease in the degree of structural integration between Years 1 and 2. However, when the content of the self-images was shifting drastically, the degree

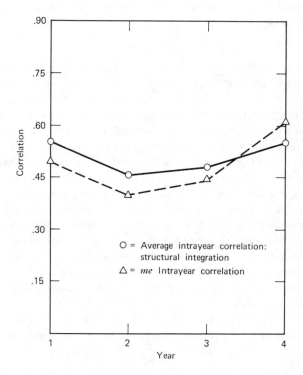

Figure 3. Intrayear correlations for Jim: structural integration.

identity. The persistence of negative interyear correlations denotes persistent *rejection* of previously esteemed features. Also of interest here is the fact that Jim's *me* shows marked variations from the patterns of the other subjects. Rather than displaying no differences, the self-image here has patterns that parallel those of Jim's other self-images. This deviation is consistent with the hypothesis suggested earlier in discussing *me*, namely, that in extreme situations even this highly conscious self-image shifts as well.

of structural integration was already increasing in Years 3 and 4. Such a combination of fluctuating temporal stability coupled with fluctuating structural integration fits the operational definition of a psychosocial moratorium.

The outcome of Jim's moratorium is progressive identity formation. This outcome appears in both sets of data on Jim. Clinically, it is apparent that his new "choices" are emerging as he now speaks of the future and goes about making his very new plans for it. The more objective and quantitative indication that this "experimentation" period has ended is in the Q-sort results. Jim's final Q-sort, taken at the close of high school, displays risesd in interyear and intrayear correlations, a finding consonant with further progressive identity formation.[12]

Lenny, a black boy. Lenny also encountered unexpected changes while in high school. He had always wanted to be a Marine, as had been his uncles and his father. He would often try on their uniforms, imagining that he too would soon be wearing one. But in his second year of high school he began to receive increasing encouragement to attend college, to take advantage of the new "opportunities" being offered to blacks. His mother and guidance counselor echoed these "suggestions." His father, as always, "stayed out of it." In the interview, Lenny remarked that a "reverse prejudice" was occurring. He felt that he was now receiving undeserved attention and support because of race, "not myself." Yet gradually he decided that the many encouragements stemmed in part, at least, from his own ability and "promise." By the end of his second year of high school, he had decided to go to college. Nonetheless, he felt that the Marines were not wholly discarded as an alternative. One could go to college and be a failure too, he observed, in speaking of the "college graduates who hung out in the playground" without work.

In contrast to Jim, Lenny's Q-sort showed rises in temporal stability during this period of changing plans for the future, as seen in Figure 4. Interyear correlations for *me* and *ma* increased between Years 2 and 4, and the average interyear

---

[12]See Year 4, Figure 3 and Years 3-4 interval, Figure 2.

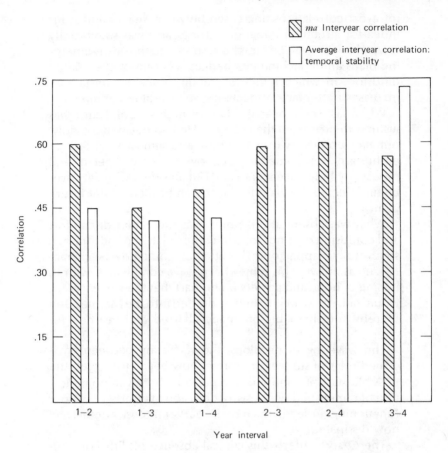

Figure 4. Interyear correlations for Lenny: temporal stability.

correlations, the indices of temporal stability, remained high
from Year 2 onward.[13] Hence a major shift in plans and long-
range goals was only minimally reflected by the temporal
stability measures. The relentless constancy in image content
noted for the blacks as a group was also evident for Lenny.

Lenny's new decisions and orientations were expressed in
other ways on the Q-sort: by significant decline in his structural integration values. As illustrated in Figure 5, his average

---

[13]Only on two self-images, *pa* and *other*, did he show any significant decline in
interyear correlations.

intrayear correlations decreased between Years 2 and 3; his index for structural integration for Year 3 was significantly lower than for Year 2.[14] Thus his response to the newly emerging plans and concomitant changes was one of attentuated diffusion, in which temporal stability remains constant or increases in the face of declining structural integration.[15]

When last seen, near the end of high school, Lenny was waiting to hear from the colleges. He had decided to apply, but had done it with a slowness and almost total lack of enthusiasm. He claimed to not know if he could "get in" and he was anything but eager to find out. Besides his anxiety over being rejected, he was also clearly ambivalent about future plans.

What was absent at this point, was the earlier discomfort and confusion over not following the "plan" he had long ago so carefully mapped out. The situation could now be characterized as showing his somewhat passive compliance, with an inkling of hope and anxiety on his part that the current plans would fail. For he might then be drafted into the Marines, and thereby return to his originally charted and yearned for course.

The wavering of decisions, the vacillating between envisioned roles of student or Marine, now began to suggest the possible start of a moratorium period. There was no indication of diminished synthesis, of diffusion of efforts, or of poor functioning. Indeed, much of the earlier intense conflict had now dissipated.

The Q-sort reflects this clinical absence of diffusion processes. Structural integration has stabilized by Year 4, but temporal stability values have continued at very high levels. Although not yet clinically apparent, it may that Lenny is handling the concomitant conflicts over this major set of changes through a return to fixed and highly related self-images.

---

[14]At $<.03$ significance level as determined by the t test. To apply the latter technique, the product-moment correlations were first converted to z values. The relevant z averages were then compared using the t test.

[15]One might argue that Lenny's decision, seemingly "forced" upon him, makes his change a radically different one from Jim's and to compare the two then makes little sense. However, this difference in "origin" of the decision change, even if it is significant, in no way explains why in the face of pressure for change Lenny demonstrated so striking a structural integration decrease, a declining synthetic function in face of a rather constant content.

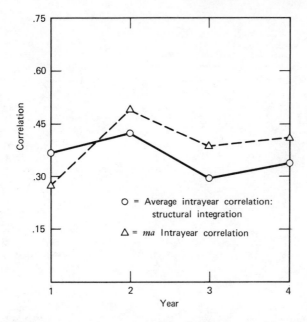

Figure 5. Intrayear correlations for Lenny: structural integration.

Clearly, neither of these all-too-brief cases fully explains the many differences and parallels in the identity processes of these two boys. The developments are obviously complex. At this point, these subjects have been introduced to exemplify the two predominant modes of identity responses to major *changes* in envisioned plans: by moratorium, which was the more characteristic white response; and by a tendency toward diffusion, a suggestion of a possibly unstable quality inherent in foreclosed identity, the characteristic black configuration.

We are left with the problem of *why* the blacks show this particular identity formation configuration. Some clinical and environmental clues to the black patterns have already been offered. However, the striking differences between the two racial groups here is by no means explained. In the next chapters the questions of determinants are explored through empirical, clinical and theoretical perspectives.

CHAPTER 6

The Roots of Identity Foreclosure:
Clinical Considerations

Introduction to the Clinical and
Theoretical Interpretations

By means of the Q-sort and in-depth interviewing over time, the identity formation of a group of black and white working-class adolescent boys was investigated. From its inception, and throughout the study, both methodological and theoretical problems were central. Methodologically, a major purpose of the research was the formulation of a definition of identity development which would render the concept amenable to objective empirical analysis. Theoretically, the research was designed to explore the interface between personality development and the sociocultural context. To approach both of these general goals, matched samples of lower-class adolescents, differing only in terms of race, were studied. The results of these studies have methodological and empirical implications.

The first set of results concerns methods. We constructed an operational definition of identity formation. This definition was based on the many discussions of Erikson[1] and of other students of identity problems.[2] Using this basic definition, we then derived operational statements for variants of ego identity development.

This type of result fulfills one of the dual purposes of the study. However, it also generates new problems: specifically, those concerning the *validity* of the operational definitions. Does reformulation of the concept into terms of the complex interactions of self-images over time accurately reflect the meaning intended in prior clinical and theoretical discussions?[3]

[1]Erikson (1946, 1950, 1956, 1968).
[2]See Dignan (1965); Keniston (1959); Lynd (1958); Prelinger (1958); Strauss (1959); Wheelis (1958).
[3]Erikson suggests that identity formation can have a "self-aspect" and an "ego-

83

One way to respond to this difficult and key question was suggested in the previous chapter. There, simultaneous clinical and Q-sort data for two subjects were compared for several periods of time. The conclusions urged by the Q-sort data were examined in light of the clinical evidence of identity processes. Further comparisons of identity development with Q-sort assessments for the same subject should be of value in approaching the problem of validity. Others have suggested means of making and comparing clinical judgments about identity development.[4] The issue here is to utilize such approaches to determine the validity of the Q-sort technique as it has been used here to study identity development. This evaluation must obviously be made if the technique is to be employed to both define and study identity formation.[5]

The second set of results is a consequence of applying the operational definitions of identity formation. Extensive study of the formal aspects of identity development of all the subjects revealed unanticipated and striking differences between the two racial groups. The black adolescents displayed the pattern of identity foreclosure. When presented with pressures to change, these same subjects responded with tendencies toward identity diffusion. In contrast, the whites expressed patterns consistent with progressive ego identity development, and when confronted with unanticipated pressures to change, they showed tendencies toward a psychosocial moratorium.

Just as the methodological results generated new problems, so too do these empirical findings. Now the problems become those of explanation. To put it simply: How are we to understand these striking racial differences in identity development? Of particular interest is the identity foreclosure of the black adolescents. This variant is one with important consequences for continued ego development. Self-limitation and *stasis* are the basic general properties that define this

aspect": "One can speak of ego identity when one discusses the ego's synthesizing power in light of its central psychosocial function, and of self-identity when the *integration of the individual's self- and social role images are under discussion.*" (1968, p. 211; emphasis added). In these terms, we have studied the *self-identity* aspect of identity formation, one of the dual aspects of the overall process.

[4] e.g. K. Keniston, "Exploratory Research on Identity," 1959, unpublished mimeo.

[5] Currently, we are using this Q-sort method *together with* clinical research interviews, an ego development measure (Loevinger, 1976) and a self-esteem measure (Coopersmith, 1967) in a longitudinal study of early adolescents and their families. Through such a multi-method investigation, we will be able to look at aspects of the construct validity of the "identity formation" measure as we systematically study theoretically related variables such as ego development and adaptive strengths (Hauser, 1979).

form of identity development. Rather than manifesting change with further experience and education, the black patterns emphasized fixed self-images, unchanging in their content or integration with one another.[6] There are disturbing implications that follow from this variant in identity formation. Minimized, for example, are the possibilities for modifications in self-image integration, in future orientations, in adaptations to unexpected occupational, educational, or social opportunities. And these forms of constriction represent but a few of the problems raised by this identity variant.

The consequences, both personal and social, of the black identity foreclosure are not in themselves a focus here. However, they cannot be taken lightly, for they are important as they stress the kinds of serious problems attendant upon such an arrest in ego development.

The discovery of black identity foreclosure leads as well to theoretically perplexing questions. Although we do not know how prevalent the pattern is, its significant prominence in this sample alone opens the problem of explanation. What, for instance, are the conditions that underlie the emergence of this identity variant *only* among the black adolescents. In this chapter and the following one, "The Roots of Identity Foreclosure," we will consider clinical and theoretical aspects of the findings.

There are two resources which we can use in our attempt to understand determinants, and implications of the black adolescents' identity foreclosure pattern. Most readily at hand are the multiple interviews in which the subjects describe themselves, their inner lives, their points of environmental success, and "soreness."[7] We have already looked at some of this data in the preceding chapter where we discussed responses to major life changes in the two racial groups. We now use the interviews to further clarify the nature of the black identity foreclosure.

In the next chapter, the second resource becomes central. Introduced are the formulations and observations of other social scientists. To what extent do these analyses contribute to an understanding of what underlies the black identity variant, identity foreclosure? Two explanatory models constructed from these other accounts are exam-

---

[6]The *content* (temporal stability) of the self-images was that aspect most markedly fixed for the black subjects, as noted in Chapter 4. The significance of this difference in degree of stasis between temporal stability and structural integration for the black boys is unclear.

[7]See C. Pierce, "Problems of the Negro Adolescent in the Next Decade," in E. Brody (Ed.), Minority Group Adolescents in the United States, 1968, p. 33.

ined closely for their relevance to the quantitive and qualitative findings of this developmental study.

In these two chapters, then, the major thrust is toward delineating the roots of the black identity variant. We first consider more detailed qualitative data about the lives of black and white boys, particularly their adolescence. Then, two models for understanding the black identity foreclosure are reviewed. The complex models are studied in light of how they handle the problem of underlying determinants: the roots of black identity foreclosure.

Our final chapters review relevant empirical and conceptual writings which have appeared since the publication of this study. Embedded in these chapters are a number of significant questions and issues raised by our findings. The most fundamental of these issues concerns the *meaning* of our original findings. We will entertain alternate modes of interpreting the black identity foreclosure pattern in light of more recent formulations which use a *difference* or *adaptive* model to explain racial difference, rather than the deficit model which was more closely tied to the explanations we offered in 1971.

THE SUBJECTS DESCRIBE THEMSELVES OVER TIME: THEMES AND ISSUES

Work

Current employment was a topic mentioned frequently and with diappointment by the black adolescents. Being less fortunate than the whites in finding steady work, the blacks held many and varied jobs, ranging from nursery school helper to grocery clerk to subscription salesman. The salesman job was despised by one black who described his daily unpleasantness when so many people "slammed doors at you." When they found work, the blacks were most often highly discontented, feeling overworked or not fully respected. Such problems were never discussed with an employer. Instead, the dissatisfied employee suddenly stopped going to work, giving the excuse of "no time" or illness on the rare occasion when he was pressed for a reason. Indicative of the work problem was the fact that the few whites whom the black subjects were able to criticize during interviews were always their employers. These were the men whom they perceived as cheating, mistrusting, and exploiting them.

Frankie, one of the black adolescents, discussed work extensively in every interview. From the beginning to the end of high school, it remained a prominent issue for him:

Work and Its Discouragements

Each year, and then at several different times during our meeting, Frankie told me of his "problem":

> I had a problem. I wanted to get a job. I never could get a job. That's all. That's the only thing that worried me, 'cause I always wanted to get a job. Then when you go look for a job you never find one ... I didn't never find one, that's all. I knew I wasn't going to find no job, but I looked anyway ...

Agencies and individual employers can never find work for Frankie. He persists in trying, responding to the many "no's" with more searching. Yet even when he finally finds a job it is dissatisfying:

> I went down to the employment office and they said they would look me up something. But see anything they want me to do I don't want to do. Like selling magazines. I don't like to sell mags... it gets disgusting and tiresome... and you meet a lot of people maybe, you get a lot of no's ... it's just disgusting.

The work is always disgusting. Frankie likes being comfortable and cared for. Work that is physically demanding is unacceptable, and employment that has rejections as an intrinsic aspect is unthinkable, for example, peddling subscriptions or other "traveling sales work." Frankie does not have a favorable evaluation of himself. Through mechanisms such as sexual bravado, denials of rejections, and grandiose visions of prosperity, he counters much of this self-degradation. He does not succeed in denying his experience with "low status" work. These moments serve to confirm his sensed inferiority. The only way for Frankie to deal with this unacceptable confirmation is to avoid it, by quitting.

Finding acceptable work is an old issue with Frankie, his friends, and many generations of his family. This history may be why he shows greater insight and familiarity with employment frustrations than virtually any other issue. Aware that he is still a high school student, not among the "bright" students, and black as well, he does not expect any but the least desirable jobs. Such an accurate appraisal does not lessen the repeating irritation and disappointment.

The current experience is also a foreboding of the future. Frankie has several clear desires for later work opportunities, but grave doubts as to their fulfillment:

I don't know exactly what kind of job I want. . . I want to be a manager of some place, that's all. [Why?] Because my father used to be a manager. . . Just have to walk around and do a little bit of work every once in a while and do my hiring and firing and junk like that. It might be kinda hard getting a job in the future. Lots of things might be difficult . . . Everything is done by a machine. It's going to be kinda easy you know they won't have a lot of people working . . . they might have machines replacing men and stuff like that . . . and a lot of people will be out of work . . . I think it might be hard because from what's going on nowadays it going to be kinda hard. [I] think about it only if I can't get a job all the time. They got, ah, they got somebody always replacing me. It's disguesting. It's a disgusting feeling.

Cogent reasons for Frankie's vague future plans, together with additional motives for his interest in the army emerge from this background of work uncertainty and unfulfillment. To fail in providing satisfactory work opportunities is the community's final insult to Frankie. He tries to deny his immense displeasure: "If I can't get one [a job] it's tough." His many direct and indirect admissions of broodings over this problem quickly repudiate such denials, however. Without any assurance of future employment, and thus usefulness, how can Frankie possibly envision the future with any clarity or security? He can daydream of coming wealth only to be rudely reminded of its unlikeliness:

I dreamed I had me a lot of money. Had me a tough car. Had me a sweet-looking girl friend. Had me a whole bunch of clothes. I had everything I ever wanted. I had a nice house, I mean those things that will never come true.

Being accepted in the army ("the service") would mean that those outside of family and friends would not totally reject Frankie, and that in the military he would at least "have something to do." Moreover, he can imagine all manners of advancement, formal recognition, and other specific signs of success. But again this appraisal of reality qualifies these dreams:

I had one good dream: I was in the service and I passed the test for lieutenant. That was good and before you know it I had been there two years, and they made me admiral I think. I don't know, a whole bunch of crazy things that don't make sense.

The white adolescents spoke less often about work. When they did discuss this topic, the comments were usually favorable. For instance, Joey, one of the white subjects, has worked at the same store for two and a half years. In this period, he has been promoted "to the floor" and received several pay raises. Although occasionally critical of his "bosses," his attitude is usually one of immense appreciation and at times frank admiration. A second white subject has had three jobs, each for almost a year. All but one of these jobs have been in the area of his choosing, photography. Some of the white boys worked only during summers and were then pleased with the jobs, although occasionally bored. Most important in the whites' descriptions and evaluations of work was the low level of disappointment, degradation, or disrespect associated with any given job. Moreover, unemployment for the white adolescents was not a catastrophe. Although they were ever alert for a good job, they did not become preoccupied with the problems of unemployment, with the need for "coins" and a "boss." Since the size of family and overall income of both groups of subjects are similar, it is unlikely that this difference in work themes is merely a reflection of economic deprivation. Nonetheless, it is obviously not wholly separate from financial considerations.

In their plans and expectations regarding future work, the two groups were again widely divergent. The difference here is well characterized in the comparison between Frankie and Joey. Frankie envisioned a limited number of possibilities for future employment, but none of these was anticipated by him as being particularly desirable. Besides being unexcited about the choices, he was most aware of their pitfalls. The risk of unemployment in these jobs was extremely high. In contrast, Joey had many plans for his future work. The plans were optimistic and numerous: "I want to get into some good business, make a lot of money, have everything turn out: good wife, kids, give them all I'm capable of."

Of the entire group of blacks, only two viewed the future as including an array of alternatives and "opportunities." Yet both of these exceptions qualified their visions with grave self-doubts as to their chances of success at any of their choices. Hence, although they differed from the other blacks in their conscious perceptions of the future, their final expectations were roughly the same: unexciting, undesirable, but not wholly intolerable work.

These two black adolescents were exceptions in several ways. One of them, Lenny, was briefly described in the last chapter. With great uncertainty, he eventually acquiesced to the urgings of his mother and teachers and applied for admission to college. Gradually, his ambition and dream of becoming a Marine was tempered. Almost simultaneously, Lenny's belief in his ultimate failure became more explicit and constant. He had seen many "college guys" in the school yard loafing all day because "they couldn't make it." It was perfectly clear to Lenny that they had wasted their time and money in going to college. He was hardly convinced that his future would be any different. Benny was one of the most unusual boys in the whole sample. From the start he was remarkably inquisitive, labile, and imaginative. His desires in both the present and future were diverse and elevated, varying from laboratory technician to Ph.D. in philosophy. He, too, questioned the attainment of *any* of these ambitions. Indeed, he almost fulfilled his underlying predictions of failure as he dropped out of school and was repeatedly unsuccessful in enrolling again. When last seen, Benny was disappointed and bitter, working full-time at the very factory jobs he had so adamantly wished never to have.

In place of the virtual sense of *predetermination* expressed by the blacks was the white's assumption of "free will". Most felt that they could "decide" what their future vocation might be. One boy had wavered for many years before and during high school. He finally chose to become a technician or "engineer." The majority of whites considered several alternative sets of plans while in high school. Joey and James typified the pattern of "conversion" from skilled labor to "professional" work. Mel and Paul were subjects who revised training programs and eventual vocations several times. On each occasion they announced a possibility previously unmentioned. In brief, work in the present and future was not seen by the whites as the dismal impasse anticipated by their black counterparts.

Heroes

In this thematic area, as in the preceding, the patterns of the black and white groups closely parallel those of Frankie and Joey. Frankie had few positive conscious or preconscious identifications. He described fleeting glimpses of his brother-in-law, uncle and southern grandfather; and even then told of their many flaws.

In place of an idealized figure stood a despised one. The theme is poignantly illustrated as Frankie describes such an "antihero":

"Pops"—A Threat and a Burden

Since childhood, "Pops" has been a feared, generally avoided, object of shame. Dislike—"hate"—of him has intensified, recently coinciding, Frankie thinks, with an increase in his father's drinking. The pattern of Frankie's despair over his father is an old one, however. Were it not for one probable screen memory and an occasional comment Frankie would succeed in presenting a thoroughly bleak view of this man and his relationship with him.

As a child, Frankie remembers that he rarely saw his father. There was a male neighbor who took him and his siblings on trips. They often wished aloud that this man were their father. Frankie's fatehr was not completely absent. Only because "Pops" taught him to "form" did Frankie finally conquer the bully who fought with him each day. This memory of father's guidance is outstanding in its isolation. All other recollections are unpleasant. Among the painful memories are "Pops" losing his "good job" because of drinking, "Pops" as a loud drunkard in the street in view of Frankie's friends, "Pops" openly flirting with street women, "Pops'" mysterious absences, and Frankie's brothers beating his father at home. Frankie has openly insulted his father's women, although he usually ignores his father in public. In fact, he tries never to be seen with him. This is no great departure from earlier habits. As a child, he felt his brothers were "Pops'" favorites, and consequently he was unhappy while with his father. Frankie is not alone in avoiding public and private encounters with "Pops." "Moms" does not go anywhere with him and she confides to Frankie, she is even unhappy in bed with him for he reeks of alcohol. Frankie has made some attempts recently to approach his father, asking him why he continually drinks. "Pops" discourages the attempts, perhaps responding to their provocative character by telling Frankie, "You wouldn't understand." Frankie's angry response to this is, "Later for you."

A puzzling question is why "Pops" remains with the family. He is subject to abuse from Frankie, and from his wife, who taunts him about his "wealthy" background which did nothing for him and his "brains" which he is wasting. He is otherwise ignored and treated like a "bum."

The motives for "Pops'" remaining at home are, with only Frankie's view at hand, barely discernible. But certain positive consequences of his being there are not so obscure. By staying home he has taught Frankie to "fight" like a man. Through talk and behavior, he has shown Frankie that sexual prowess is a critical dimension of manhood. For although Frankie is bitter about "Pops'" extramarital sexuality, it also represents one of the few ways in which Frankie is proud to resemble "Pops."

On the negative side are failure, disrespect, deprivation, and much anger. "Pops" is a man "who could have made it." He is not, to Frankie, stupid or incompetent. For some unknown reason, he has perpetually "messed up." Even with all the advantages of supposed wealth and intelligence he has failed. The unhappy career of his father remains an enigma to Frankie. Yet whatever his confusions about the "why" of his father's fate, the "what" is only too obvious: a fallen man. And whatever failings he cannot himself perceive, he is forever told of by "Moms": the paychecks wasted on drink, her shame of being with her husband, her surprise that he is not yet "sick" and her desire to separate from him. There is some suggestion that "Moms" may expect irresponsibility and failure from all men. Certainly her repeated injunctions to Frankie not to drop out of school, not to get "in trouble," not to marry young, hint at a pessimistic view of *any man*. There are few desirable goals that she depicts for Frankie. Perhaps if he can avoid the pitfalls inherent in being a man he will be told of its potentialities and rewards. Her expectations are reminiscent, though possibly less severe, of those held by some black New Orleans' "matriarchs":

> Men are all alike. Men with six or seven kids at home, sitting at the bar fooling around, spending money. I don't have anything to do with men like that. I like to sit there and watch them. I have no sympathy for them.[8]

"Moms'" actions and attitudes toward "Pops" have been justified to Frankie in terms of his irresponsibility and general unworthiness. Frankie has witnessed "Moms" managing the family and home, while "Pops" remains the "burden," barely tolerated, heartily disliked, and scorned.

An important question is whether or not these many years of observations have encouraged Frankie to view men and thereby himself as always second in command to women, inevitably less reliable and weaker. While such specific information is not available, we do know much about the closely related problem of Frankie's confused sexual identifications.

Until the age of eight Frankie enjoyed playing with dolls, stopping only when he learned that "boys don't do that." An important element of this play was tearing the dolls apart. Aside from dolls, children of the opposite sex were favored companions. Throughout childhood, and currently, Frankie's happiest times have been with women and girls, whether talking or have sexual adventures with them. His hair has represented a sexual issue. While interested in "processing it," he has rejected the plan, in large part because "it's like going to the beauty parlor." The impetus for a fight Frankie began on the bus came when the opponent "hugged me like a girl." A more explicit homosexual incident evoked great disgust and violence in Frankie toward his eager seducer. Further sexual preoccupations are revealed in his frequent and lengthy narratives of sexual conquests, describing in elaborate detail virile adventures and successes. Whatever feminine tendencies the above adventures and battles may be responses to are more

[8]Rohrer and Edmonson (1960), p. 129.

directly seen in picture-drawing in which Frankie consistently draws a woman as his first "person."

There are many grounds for Frankie's seeming doubts about his masculinity. In addition to the degraded pictures of his father drawn by "Moms," his experience with "Pops" has been minimal and often disturbing. When with his father he did not feel wanted or liked as much as his two older brothers. Now as an adolescent he is either rejected by his father in the few approaches Frankie makes; or he rejects father verbally—"Later for you,"—or physically when Frankie beats the intoxicated, irritating man.

While contact with "Pops" has been limited, and then adverse, experience with "Moms" is of an entirely different nature. Almost all of Frankie's time at home is spent with her and is usually gratifying. If not with "Moms" Frankie frequently is with one of several favorite aunts, or grandmother, or his older sister. There are an abundance of admired females, but males are few and undesirable: two brothers who "messed up" and a father who continually exemplifies failure. The few more "positive" males are either distant or cautiously admired: a "bank president" uncle in the South, a black football star and, more recently, an uncle who supported and helped provoke several of Frankie's fights during a visit to the South.

In view of this history, it is not surprising that Frankie expresses much anger and resentment toward his father. "I hate him," "I wish he'd get out," are typical spontaneous and heated remarks in the interviews. These feelings are probably embarrassing as well, at least when expressed to a white adult, for Frankie waited two years before mentioning them in our interviews. There are many ways in which "Pops" has "done Frankie wrong." Though he did not desert the family, as many of Frankie's friends' fathers did, it was almost the same. Feeling deprived of father was a repeated experience. In addition to feeling abandoned, Frankie witnessed and at times participated in the drunken assaults, which he could actively respond to only many years later. In sum, "Pops" is an inconstant and threatening figure. It is unpredictable when and how long he will be home, whether he will be violent or peaceful, kind or rejecting.

Of interest is the fact that Frankie has up to now not chosen "the gang" and organized delinquency as a solution to the absence of adequate masculine models and symbols. In this respect his nondelinquent older brothers and the occasional yet available exposures to favored uncles probably been important. Furthermore, the mere presence, no matter how unstable, of his father at home who is likely to have been critical in providing an image of manhood as well as tempering distortions of men that "the women" may have created. In terms of the future, there are several potential directions Frankie may take. One possibility is minor delinquency and petty crime. He has on occasion shown interest in this; police have frequently supported images of Frankie as a "hoodlum," as tensions with "Pops" have become increasingly unacceptable in the last two years. Homosexuality does not seem to be an acceptable solution, and withdrawal from all social contacts is also intolerable, as Frankie shows an everpresent and strong wish to be accepted. One solution, which Frankie himself offers with considerable ambivalence is "the service." A short period in the armed forces may provide Frankie with both a respite from his immediate family and many new, perhaps more acceptable, male models and symbols.

Frankie is a specific instance of the general trend observed in Chapter 5. Most of the blacks frankly stated that they had no heroes, no one they wanted to resemble now; they wished, "just to be myself." Occasionally uncles, teachers, prominent black athletes, and businessmen were selected as ideal figures. Black responses were plentiful when they considered antiheroes, those people whom they wished above all not to resemble. The subjects then spoke of bums, beggars, drunkards, gangsters, and thieves. For several boys—such as Frankie— the most important antimodel was father. The possibility that at least some of these emphatically detested figures were strong unconscious identifications was suspected for several subjects. The high correlations on the Q-sort between *me* and *pa* for blacks further supports this speculation.[9]

The black subject's deep ambivalence about his father is clearly shown by Jerome. Briefly described in the last chapter, Jerome, for many years, had the ambition to become a physician. Despite an unremarkable academic performance and an unahppy school adjustment, this wish remained strong. His father, a postal clerk, was rarely mentioned in any of the interviews. When Jerome did speak of him, it was always in an uneasy, self-conscious way. At the time, Jerome's mother was completing her training at the state teacher's college. She was clearly the principal adviser and motivator in Jerome's thoughts about college and professional training. A related issue for Jerome was his increasing discomfort in remaining in the study. This culminated in his finally telling me that he could no longer come to the interviews since he disliked talking about himself. While in the study he had become highly suspicious of my motives for speaking with him, more than once telling me that we were studying "Negro families" like the

[9]The topic of "heroes" or the more general one of role models is discussed in the next three chapters in terms of theoretical implications as well as recent empirical evidence.

studies in his mother's textbooks. In addition to his anxiety over a white, or anyone, peering into a potentially unstable family arrangement, it is likely that his withdrawal was a response to the many interviews with an adult male. He was having to speak and introspect with a man—and then a white one—who reminded him of attributes not found in his father. The conflicts aroused by the study may well have been beyond his capacity or desire to tolerate.

Benny represented another variation of the father theme. He openly and repeatedly berated his "stubborn" and "mean" father. The latter, separated from the family for many years, was seen by Benny as an outcast, and undesirable. Benny visits him only for "cash." However, money seems to be in large part a rationalization for the visits. Benny often hesitantly expressed the less critical and affectionate feelings he held toward his father as he recalled the talks and "fun" he had with his "old man."

In discussing Frankie, we noted how the status of mothers for the blacks is very different. Mothers were always described with the strongest superlatives. Many of these women sounded exactly like Frankie's "Moms." They were advisers, providers, and "bosses." Any problems from the most trivial to the most profound could be solved by them. They had major authority roles in the lives of their sons. Subjects were often worried about "Moms" and her reaction to their behavior, school failure, or unemployment. To speak of heroes for the black adolescents is probably incorrect. They admired, respected and revered a heroine, "Moms."

There were many many heroes for the whites. The contrast between black and white adolescents here is richly elaborated as we compare Joey and Frankie.

"Fine and Successful": An Abundance of Heroes

Frankie could easily describe those men and boys he wished not to emulate. Finding desirable heroes was far more difficult for him, although when pressed he could recall a few men he admired. For Joey the problem was reversed. He had always been surrrounded by relatives and others who were "the best," whom he wanted to "be like." These idealized figures are great inspirations yet almost always unreachable. The consequence of this was chronic dissatisfaction. It was a dissatisfaction without the element of despair; for Joey retains the image of what he can be, if he "tries harder." Frankie, however, expresses a self-dissatisfaction coupled with despair. For Frankie the visions are of what he can and will be if he is not careful. The majority of images of the future for him are foreboding. Only a small number of the men whom he knows or is related to have "made it." The primary problem for Frankie is to avoid becoming like the negative models; the images of those who have "succeeded" are neither vivid nor plentiful ones. All of this, of course, is in terms of conscious or pre-conscious emulation. There is every reason to believe that unconsciously Frankie has very strong and basic identifications with precisely the men whom he now consciously shuns. These are the men such as his father and others he has known and heard of all his life.

There are but few people whom Joey wishes to avoid resembling. His antiheroes, small in number, all have certain features in common. Their failing is associated either with emotional weaknesses and conflicts, or with work. He does not want to resemble one of his aunts who is a person "with a very bad disposition. She's very slow, takes a long time to get started and gets aggravated if you rush her. She doesn't understand if you rush her." Another kind of emotional problem, far more despicable, is that of rampant aggression and impulsivity:

> The Boston Strangler he killed a lot of people, that guy. If he gets caught boy it's all over. They only kill 'em once. But I think it's gonna' be painful, 'cause they gonna' wanna', like the guy that assassinated Kennedy. I'd hate to be in his shoes when he got caught, Oswald . . . He woulda' died a slow death, Oswald, even if they let him rot in prison. What people would say and what people would do to me if they ever saw me [if I were like him].

> That guy's weird [the Boston Strangler] . . . Well, when he gets caught he knows he's gonna' die for the one murder he committed. But just think if he had all those lives to live he'd die all those times. Fifteen times and uh he's gotta' have that on his conscience. He's gotta' think about it once in a while, how he killed the girl . . . He must have nightmares. Even if he didn't have those nightmares it's just when he's walking down the street or something there's gotta' be somebody that resembles someone, like say somebody has the same dress on that the lady he killed had on. So he just looks at her ya' know and he starts getting nervous and confused. And it brings back memories of what he did, thinking of how he killed and everything.

In Joey's descriptions of these antiheroes, are embedded the greatest fears: guilt, death, and violence—about which he is highly conflicted as well. Joey is typical of the white subjects in the importance he gives to conscience and guilt. Although there was variation on this issue with Joey representing one extreme, the entire white group nonetheless differed markedly from the blacks in this respect. For the blacks the overriding emotions and conflicts centered on rejection and inferiority as illustrated by Frankie.

A second feature shared by Joey's negative heroes is that of degrading work. Here, there is a surface similarity to Frankie, to whom the *wrong* kind of work was a critical problem. Joey's objection though was almost a fine distinction alongside of Frankie's gross rejection of certain jobs for their humiliating features. Joey's concern was about fair compensation and adequate working conditions rather than despicable forms of work. It was a *union demand*, not a demand for a *union* and a *new image*:

> [I wouldn't want to be like] my mother now. She's a cashier and has these long hours so sometimes we never know if we're gonna' see her or not. She comes home late at night after leaving early in the morning. I wouldn't want to be like my father in the future because his job has back-breaking hours. He works long hours and takes a lot of money out [for taxes].

And Joey does not want to be like his teacher:

> ... The kids give them a hard time all day. It's a hard job, a lot of kids. They put a bad mark without asking you to be quiet. They suspend the kids...

Joey focuses on work describing people and life styles he would not like to resemble. Work had undoubtedly been a primary target of the many gripes and bitter comments he had heard from respected adults.

As far as people who are attractive models for Joey, there are more than 30 men whom he admires and in various ways wants to emulate. This number increases each year. Foremost among these virtual idols are Joey's relatives. For great-uncles and a cousin he has unqualified praise.

> I wanna' be like my great-uncle who helps with the church. He's very famous, very good and well known. He'll do anything he can for people; if he can't he'll have others do it. He helps parish families' juveniles in court ... They wanted to make a movie about him. He worked his way through col-

lege. He's mayor of his village. I want ta' be like my uncle who's in the construction business. He's one of the best masons in the city. He helped build this building. Everybody knows him. He's famous.

His older cousin is a more immediate hero.

My cousin was prom chairman last year and uh he had to turn away a few hundred couples because it was so crowded ... I guess that was a big responsibility and everyone looked up to him. I asked him if I was elected if he would help me ...

The position Joey takes toward his father is more complex. For many years prior to high school Joey wanted to do the same type of work as his father. To his disappointment he was discouraged:

I thought of working near my father as a welder but he talked me out of that. He didn't want me to work those hours. He worked twenty-four hours when the railroad was active. I guess everyone wants to be like their father because they admire him.

Since then he has been critical of his father's work, but also very pleased at each opportunity he had to work with him:

I like working better [than at other jobs] with him [father]. Though it's dirty, it's nice to see the room change from empty to filled with sinks and stuff like working with a man's hands instead of machines ... When we were all through he would say "carry the tools back to the car." He would pick up these big pipes. He was all right. He was good to me you know.

Joey has many times enthusiastically told of a summer when he worked with his father, who is a plumber "on the side." He again worked with him after his family moved. This time he was two years older and less patient with the authoritarian regime:

He's [father] funny. If you start something you gotta' finish it. He won't put it off. We screened in our porch, screen windows ... and I said "but Dad, I'll do it this way." And he said, "You don't want to do it my way?" And I said, "No." He said "You're going to do it anyways ..." Now the windows are all in place ... He's the authority.

Despite the apparent problems with this firm ruler, Joey makes it clear through this and subsequent descriptions of their work together that such times are special ones.

When moody, Joey has been told "you are just like your father." In his friendliness and "popularity" at school the resemblance to his father is also apparent:

> ...He comes home in a bad mood or something but my father is very friendly. We used to go out riding with him and we would see somebody and he would go around the block... And he would start talking and talking and they wouldn't stop.

Though he may behave in similar ways, Joey is painfully aware of the fact that he does not physically resemble his father. Joey's shame of his body and short stature is clearer in light of his description of "Pop." "He's sort of big, bigger than you [5'6" interviewer] much ... About five ten, 230, 40 maybe [pounds] solid. Big guy."

During high school, Joey has begun to look outside his immediate family for models. He has not, to any extent rejected his idols in his family. They are supplemented and "updated," but remain the foundation upon which all later heroes rest. Two recent figures were high school "athlete-scholars," who faced a choice between physical or mental prowess, a decision Joey is also concerned with. At the time they decided to go to college, Joey announced to me that he too wanted to go on to "college." A third of these new heroes was the astronaut, Edward White. What impressed Joey most was his act of rebellion, his refusal to return to the space ship after having stepped outside. Choosing this characteristic is consistent with Joey's tempered forays into rebellion throughout the three years of high school. White's rebellion was not a flagrant violation:

> I think that he [White] was too thrilled to think of getting back in. That's why he didn't think anything of the time that elapsed from the ten minutes he was supposed to be out ... I would probably have stayed out as long, if I could've, if the whole United States wasn't yelling at me to get back in (chuckles), I would've stayed out. But he did the right thing ... He just proved that many can take that endurance in space up there.

There are many other heroes, ranging from Gary Cooper, to new singing groups, to President Kennedy and the local mayor. Each has something Joey wants to have, be it money, fame, or "ruggedness." Such a profusion of ideals is not confusing to him. Somehow, they are all related to various themes and issues in Joey's current life and do not in any way lead toward severe diffusion of his self-images. Both because of their number and variety, these many heroes give increased impetus to Joey's efforts to self-improvement.

Joey is typical of the white adolescents. They described numerous heroes, and these heroes appeared to have much influence on the boys' decisions about the future. Joey exemplified this in his decision for "college." Two other boys decided to become teachers during a year in which they were taught by two men, men whom they were openly fond of and admired.

The Future: Limitations and Successes

The future had very different significance for each subject. His skin color was a major determinant of its significance. The black, for instance, saw a future that mirrored the present as he knew it. He might repeat the persistent series of discouragements, or he might live a life resembling the style of adults who now surrounded him, the adults who "never made it." Among these adults were even those who—as Lenny put it—despite "education" were still "on the street corner." The images of storefront drunks and deserting fathers were also, of course, close at hand. It did not require eliciting less conscious fantasies to bring forth images of failure.[9] There are some pictures of the future that are *less dismal* than others. Frankie, for example, imagines a bleak but not wholly despondent period that will follow the completion of high school. His tortured uncertainty and pessimism reflect an outlook and style characteristic of the black subjects.

The Future: The Service and Unwanted Marriage

Among the few inspiring figures in Frankie's surroundings have been those in uniforms of the armed services. They possess unequivocal signs of belonging to an honored group of men, at once responsible and important. Frankie is well aware of their influence on his future plans:

I'm going into the service. I'm going into the Air Force or Marines or one of them ... When you see a real man from the

---

[10]This is similar to the point made by W.H. Grier and P.M. Cobb in *Black Rage*, 1968, regarding the role of abundant black prostitutes and the degraded self-image of one of their female black patients.

Air Force or Marines from anywhere around here, I mean you want to go in it. You say you are going to do it when you grow up, so I just did that when I was little ... it comes up every once in a while. Then I say, that's where I'm going to go, something like that. I want to wear a uniform 'cause they make you look nice, make you look neat ... I want to be important.

Frankie has no other plans about the service. For many years he has known he will join after high school. Yet whenever he reconsiders, the idea becomes more tenuous. A particular stumbling block has been the extent of his commitment. The military in part represents to Frankie, as it does to several of the subjects, a psychosocial moratorium. However, to each of the boys there are specific expectations and anxieties about this desired period. Frankie does not want it to mean a "quitting," or permanent departure from the community, or more important, from his family: "I'm not just going into the service and give up and pass my time away." The specific duration of service is a major point of anxiety and confusion:

> I might not want to wait five years, no I might not want to wait ten years. I'll probably wait five years and then get out [of the service] ... If I stay there for quite a while, if I stay there for about four or five years and then like it, then I'll tell them I want to go ten years....You might want to get married or something like that when you get out you know. That's what I was thinking, 'cause you might now want to stay in there all your life.

Besides his worry of looking like a "quitter," Frankie worries that the service may rekindle his authority problems: "I might not like getting bossed around. You have to take orders." Yet there are potential immediate gratifications: "Best thing I like about it [the service] ... I guess the food. "Cause they say you can eat all you want ... something like that."

All things considered, the military offers a threatening set of problems. He will be with only men and mostly whites at that for much time; he will be away from his home and mother. There are inherent authority as well as probably homosexual conflicts. The pleasure and general reassurance of female companionship will largely prohibited. Each year Frankie becomes more tentative about the service.

Frankie's thoughts about a future marriage are linked to plans for shortening or changing his military career. He rarely speaks spontaneously of marriage aside from discussions about the armed forces. When he does entertain the marriage topic it is in distress, in response to fears of "messing up," being forced to marry. Marriage would consist of a series of cautious compromises with potentially unworthy women:

> I wouldn't do my wife wrong. I would stay home and I would just have to give up a lot of things. I would work and every-thing. You know support the family and jive . . . Just as long as she don't do me no wrong or nothing like that I'm going to go along with her . . . do all the important jive that they say when you get married . . . In sickness and all that jive. I mean I'll help out. I'll do my part as the husband . . . Anyhow I'm not think-ing about getting married. I like to have fun. If you get married you can have fun, but you have to get out of school and work and all that junk. It's disgusting.

The few pleasant fantasies about marriage display Frankie's intense dependency wishes, now to be fulfilled through wife and money:

> I'll get a job before I get married. 'Cause I'm going to have a lot of money saved up anyway . . . Make arrangements you know. I'll get married and all like that. Just lay up and take it easy for the rest of my life. Ah think me a lot of weird things, boy. I swear I'll make me a good ole' time.

The future and absent heroes have much to do with one another for the black adolescents. Related to the missing heroes and abundant *real* failures are the dismal and tenuous images of the future. There is little—as Frankie so emphasizes—that is desirable and obtainable in the future.

The whites "looked forward" to graduating from high school, to the new opportunities and possibilities they envisioned as now becoming available. For Joey the future had very specific meanings:

The Future: Business and Money

Joey's visions of the future are closely connected with those of work. Rarely does he think of one of these topics apart from the other. The unusual times are when he thinks of money and the future, but even here the link with work is almost a direct one. He has many plans and wishes for future occupations. Both in breadth and in number of conceived possibilities he is once more so strikingly different from Frankie.

Chronologically, Joey first wanted to be an airplane pilot. This was followed by wishes for the priesthood while in junior high school. Soon after, his thoughts turned to the idea of being a welder, "working near my father." Welding and preaching were both discouraged by his father and mother, respectively. He was "talked out of being a priest" because of "too long a schooling." As for welding, his father "talked me out of that; he didn't want me to work those hours he worked when the railroad was active." By the start of high school, having been discouraged from both mentally and physically demanding jobs, Joey had narrowed the choice down to business manager or construction worker. Being a businessman would depend upon college, which was an uncertain but hoped-for goal. What he wanted from his future work incorporated important wishes: "If I have a good position I'd like it, like foreman, manager. It would be good paying and people would look up to you, make you feel like you're a father, ask you for help."

His general plans of business or construction work remained stable through the high school years. Changes within the framework included a refinement of the business category, the category Joey favored anyway. He began to think about IBM training and electronics. The goals or "dreams" were the same. Joey did not think that there would ever be reason to alter them:

> ... make a lot of money, have everything turn out: good wife, kids, give them all I'm capable of. I want to see other parts of the world, of the country. I want the business to be one where I'd have to travel a lot, see how the different businesses are making out.

There were many people who directly influenced Joey's ideas about future work. His father virtually commanded him to become a business man as he ordered him to not take a "hard" job like his: "He said I should be a 'white collar worker.' I told him I'd try to go to college and become an accountant." Others who supported Joey's upwardly mobile plans for the future were his "successful" uncles, and a cousin, a doctor who "helped addicts get off dope."

As opposed to Frankie, Joey described but minimal fears of future failure, of unemployment or of financial indebtedness. In part, this difference is conditioned by Joey's active use of denial. To some extent, even if he were concerned about the future he would withhold this from himself as well as the interviewer. But there were a sufficiently high number of interviews and samples of Joey's behavior to have at least suggested some anxiety regarding future failure, if such a latent theme were present. What fears he had concerned the problem of his training after high school. Holding a relatively mediocre pre-high school record, and average high school work, Joey had grave doubts about "making it." To be accepted for further education was important for his current status vis-a-vis his college-bound group of friends and for realization of his "businessman" plans. Conflict over this became increasingly intense as he neared his senior year. His fears of rejection by college suddenly disappeared with an unexpected rise in his grades; and with what he read as unequivocal encouragement by his guidance counselor of his ability to pursue further education. Still unsure of his qualifications, this confirmation by his counselor became most important.

Further progress in Joey's image as a future businessman was accelerated by achievements in his current job. The most recent promotion placed Joey "on the floor." He is now a cashier, wearing a tie and coat each day and feeling more like a businessman than ever before. The transition from high school student to future manager is already in progress. In fact, at this point it might be reasonable to predict that in 20 years Joey will be a "supermarket manager," with a large family and a home in the suburbs, encouraging his son, Joey Jr., to be a professional, such as a doctor, or lawyer, not just a "white collar worker."

DEGRADATION AND DIMINISHING INFERIORITY: SELF-IMAGES

Interwoven throughout the preceding narratives—and through the years of interviews—were the blacks' degraded self-estimates, their unremitting belittlement of themselves. Frequently, the scathing judgments were implicit in other descriptions, in other discouraging episodes. Sometimes the self-opinions were explicit, painfully obvious to the subject and shared with the interviewer.

The themes of worthlessness, undesirability, and uselessness recurred in many contexts. There were no black adolescents for whom these themes were subtle. For the whites such topics when present seemed minor; worthlessness did not assume the same unmistakable prominence. A white adolescent would question his value most often at moments of "self-doubt." Even then the questions would be modified by a fundamentally optimistic belief that things would always "get better." Once again, the lives of Frankie and Joey reflect the poignant divergence between the black and white adolescent in this area of self-degradation.

One response Frankie had to inferiority was through his personal philosophy.

An Ideology of Money

While marriage and the military are highly tenuous desires, money is the most certain and tenacious of all Frankie's goals. To have money is to be assured of being able to "lay up" (be comfortable). To have money means no longer any sense of inferiority when with those who have "coins."

> Some boys that have money you know they think if you never have no money you know they think you was going to beg them for money. ... They don't bother me none ... If they feel like giving me coins to buy something I mean I don't mind myself. Long as they don't try and say I beg all the time. Which I do not beg.

The magic substance, money, and the material possessions it allows, are the answer to all problems. The multiple fantasies of "success" so constantly drawn by television, radio, film, and the street have obviously had their impact here. The intensity and almost complete one-sidedness of Frankie's money wishes suggests that they are something more than reflection of an admittedly dominant cultural theme. To be "secure" and to be upwardly mobile, for example, having "a house in a new neighborhood," are important consequences of having money. Most of all, though, Frankie's envisioned utopia would consist of protection from degraded self-images and from the multiple threats represented by other people. Frankie would be surrounded by things of all kinds and be at once elevated and insulated:

> ... I'd wish for some money. That's all I would need. If I had some money then I could get me just about anything. I'd want then ... buy me a new home, buy me a car, buy me some cars, buy a lot of things, buy me some clothes. If that could happen right now I'd wish for that stuff. Buy me a new hi-fi, hmm, buy me a new tape recorder, buy me two or three new tape recorders, four or five amplifiers, three or four record players, all the records I could buy. Ah, I mean I'd have my house fixed up real sharp. Wouldn't that be a good idea? Buy me some new furniture for myself. Fix up the den.

Occasionally a wife and/or family is attached to the reveries of money, but these people are decidedly peripheral to the immensely gratifying dollar dreams. It is as if such people are added because "you're supposed to." The panacea is money. With it Frankie can cloak all uncertainty and protect himself from all opponents, finally becoming "king."

Other elements of Frankie's ideology are rarely so explicit. The world consists of "haves" and "have-nots" and, less easily expressed, whites and blacks. Frankie's few but significant reflections on "having this kind of hair" and his very conscious experiences of racial rejection together with his awareness of money make it clear that he is doubly degraded: he is both black and a "have-not." Consciously, Frankie believes that in large measure one can overcome inferiority through riches. The catch is that only by magic or unexpected gifts can these riches be acquired. The solution is not to be found in working; for work presents many problems and pitfalls. It most certainly cannot bring many of the possessions one needs for success.

Aside from money, Frankie is interested in action and "thrills." Women and athletics are major sources of these satisfactions. In general, women are to be exploited financially and sexually. There is a limit to the application of this dictum. Frankie frequently feels guilty for "taking advantage" of girls. In fact, one of Frankie's few moral precepts is ultimate respect for females. Toward men such a commitment is not evident. All other of Frankie's moral principles and judgments are apparently derived from the basic commandment, "Never disobey your mother." Trouble with "cops," problems in school, or early marriage would be judged wrong because "Moms" is against them.

Frankie has no formulated opinions or interests in anything that is not of immediate consequence to him. He finds international, national, or community affairs of no importance. As observed above, he is indirectly preoccupied with social class. Racial discrimination is ignored until he is confronted with personal consequences of it. He then responds to the one specific problem at hand. Though living in the midst of active civil rights movements, Frankie has no wish to either join or learn about them. The current "civil rights" environment has undoubtedly influenced his recent protests against bigotry at school, but any conscious preoccupations about racial issues are for Frankie restricted to the Muslims. Their beliefs in separation, violence, and idolatry are spoofed at and resented by Frankie. He emphatically rejects the movement and denies any interest in it. Moreover, as if suggesting the potential temptation this group has for him, Frankie notes that he will be happy when this group disintegrates.

Frankie has, then, a relatively simple ideology. He has scant conscious concern with his past either in itself or as an influence on his present life. His future is but thinly outlined and then usually only upon request. Political or moral devotions and preoccupations are of little interest to Frankie. Ideas, conscious beliefs, and abstractions are for others. For Frankie the immediate "kicks" are what count. Even the barest planning or "arrangements" are painful and are only occasionally attempted. While ideology qua ideology has not been a crucial concern, the issue of work has been. It is in this realm that society's appraisal of him and his relationship to the community is expressed. These messages—as depicted earlier—have been uniformly discouraging.

Included in themes of work, heroes, and the future have been many indications of Frankie's self-image and its vicissitudes. Though it was not always intended, many of Frankie's tests and interviews have portrayed the panorama of self-evaluations and visions that he presents and in general *dislikes*.

Of Hoodlums and Kings

Frankie has used a minimum of ten names in describing himself over the years of our meetings. The significance of each name, its conscious and unconscious meanings, obviously vary considerably. The mere plentitude of epithets and the richness of images they convey is in itself testimony to Frankie's present and quite possibly lasting conflicts over his self-images and identity. In the following discussion these names or "self-labels" are described and their meanings elaborated.

1. *Good boy.* "I'm a good boy, I think," is a statement usually appended to Frankie's descriptions of battles at school or with the police. It is said with obvious equivocation, at times with much teasing, as if he were a young boy having just been scolded and now telling of his supposed virtues.

2. *King.* Frankie wants to be "king" and to be called such. It means to be popular and powerful among all people. His popularity is uncertain, always contingent upon performance in school, on the street or, more recently with white girl friends. The aspect of power is in part one of physical strength and Frankie claims respect for this as he "protects" black sophomores at school. However, the other meaning of power rests on money and prestige. These, as we have seen, are Frankie's dearest longings and the ones he senses are least likely to be attained.

3. *Playboy.* This is a generally valued appellation, with aggressive and masculine overtones. Throughout the years of interviews, he has shown preoccupation with "babes." In all relations with them, the goal is to "get them down" (have sexual intercourse). As Frankie repeatedly emphasizes, he had no great difficulty succeeding in this quest:

I always get involved with a girl. That's the only way I have fun.
Yeah man, ah always have me some fun with them ... Ah do
everything ... That babe is crazy about me. I don't know
what's wrong with her. I'm a mess, right? ... I got this strong
game. I know how to run a game on a babe, make 'em like
you ... All the babes love me too much. I'm a mess. I game
'em all down.

His use of the name "playboy" increased considerably in the last two
years of the study, possibly reflecting Frankie's increased uncertainty
over his own masculinity. Although the reasons for Frankie's greater
reference to this name are unclear, the passages leave little doubt of
the self-deprecation that accompanies it. It is as if to once again
remind all concerned that he is essentially "a mess," in case we forgot
while listening to his accomplishments.

4. *Bum.* Frankie says he wants least of all to become a "bum," as are
the drunken men sprawled daily on "the avenue." His father in part
represents such a person to him, thereby heightening Frankie's fears
of this outcome for himself. Although a heavy smoker, he avoids all
alcohol, vowing not to "take after Pops" in any way, especially in
drinking.

5. *Lazy.* On the road to being a bum is to be lazy, and Frankie
frequently thinks of himself as lazy in relation to work and his dislike of
most unskilled jobs. In addition to his father, he has his older brother
as a model of "laziness."

6. *Hoodlum.* Frankie once remarked about his girl friend's mother
who did not like him: "She must have thought I looked like a hood or
something ... asking me all sorts of questions as if I was doing something wrong all the time." Being thought of as a hoodlum by his
mother, if the perception is accurate, continues a tendency begun
many years ago by several groups of policemen as they detained and
interrogated him for walking on the local university's campus. It is
intermittently reactivated. Now this view of Frankie is spreading to
older women, as well as to people whom he usually considers "right."
The dismay and uncertainty that this change engendered were apparent as Frankie asked me, following his story about his girl's mother, "I
don't look like a hoodlum, do I?"

7. *Smart.* This word has two different meanings for Frankie. One is
his retort to teachers' and others' low estimate of his intelligence: "I'm
intelligent. I'm smart." He needs much support on this point; he
usually makes it clear he believes himself to be stupid. In fact, one of
the functions the three and one-half years of interviews unwittingly
served for Frankie was to encourage him along these lines of hesitantly

questioning this belief. He has been able to talk with a professional, a "doctor," and to participate in "research". The other meaning of smart is "wise guy." Especially with men, Frankie is often smart in this way, acting in a provocative and frustrating manner. One of his main aggressive outlets is through "being smart."

8. *Crazy.* During one of our conversations Frankie suddenly told me, "I thought you were going to say I was crazy." He has on other occasions spoken of "seeing things," quickly adding that "I don't see nothing that might make you go crazy." There is good reason for Frankie to entertain such doubts. Episodes of aggressive outbursts have increased recently, disappointments and disturbances with "Pops" are now more acute; rejections when he visited relatives in the South were bewildering and resented. Less easily specified is the diffuse uncertainty Frankie is showing about his manhood and future. There are no data to suggest that he is experiencing any delusions or hallucinations. More likely, it is his diminished control over his aggressions and more rapid mood changes that imply "craziness" to Frankie. It is of course uncertain how adolescent-specific these changes are. There is indeed the possibility that these changes and the ensuing doubts of his sanity will persist throughout an erratic adulthood.

9. *Blue boy.* Recently, Frankie has spoken of racial prejudice and rejections. He usually handles these in the way he treats other very troublesome conflicts: "I try to forget them [those times when discriminated against]. If you think about them too much you go crazy." Sometimes, as in an episode in which he perceived flagrant bigotry in his gym teacher, Frankie directly confronts the antagonist. Then he does not as readily deny the impact of the event or subsequent feelings. The discomfort following the latter incident is still great, and Frankie "tries not to think about it."

Even more troublesome is Frankie's specific skin color which is a dark brown. Much has been written of the status significance of the varying skin colors within the black community. Frankie tells several stories about his family perusing their group photographs with guests. Much laughter always ensues because Frankie is "dark," while all the other children, except his older sister, are "light." Frankie, though, emphatically denies that color differences among blacks are of any importance to him: "It don't bother me." In choice of girls, as in self-evaluation, skin color—he insists—is irrelevant. His older brother, foolishly in Frankie's eyes, "likes light better."

To further complicate any inferiority he may been because of his dark skin is the fact that Frankie's father is also dark. Frankie is preoccupied with not wanting to resemble this degraded man in any possible way. Yet here he must face the most obvious resemblance of all. It is not surprising, then, that Frankie engages in denial around this conflict-laden area. Although he suspects many personal weaknesses, to him his most critical is that of his color. His siblings for many years could not be stopped from reminding him of his unwanted skin color as they taunted him with "blue boy" (dark skin). The label continues now with white classmates "in jest" shouting "blue boy."

10. *Up the road.* Some of Frankie's most difficult experiences occurred in the South, following sophomore year of high school, when he suddenly realized that he was a northern black and estranged in many way from southern blacks. This would probably not have been so disturbing were it not that Frankie's immediate ancestors, and oldest sister, were also southerners. The blacks he met were, on the one hand, "stupid" and "primitive" people to be shunned. On the other hand, they somehow included his own relatives and in part himself. If he were to reject or downgrade southern blacks did that then mean his own origins were lowly and to be ashamed of? The rejections and abuse Frankie received from southern black adolescents make it clear that Frankie was an alien. This left him in an extremely awkward position, for he had not previously thought of himself as a northern black. With this new concept, additional dimensions to his self-image emerged. To be "from up the road" means, among other things, to be less like deprived, segregated blacks. It also—and here is where the confusion lies—means less pure, more like "white." To be similar to the white man is a wish which for Frankie and many other blacks is fraught with ambivalence. Witness, for instance, Black Muslims and Frankie's angry protests against them. To learn that he was a northern black was probably helpful in terms of self-definition. Yet the disturbance generated by the experience was intense and lingered on. The distress from the summer was soon manifest when Frankie assaulted a new boy on the school bus, a southern black.

Such an ever-expanding catalogue of self-derision was not evident for the whites. Rather, as Joey's development illustrates, the process was in the opposite direction. The self-derision themes were there, but becoming less and less confirmed by both Joey and his surroundings:

An Ever-Diminishing Inferiority

First and foremost in Joey's self-image is his body concept: he is "short" and "skinny":

My arms are skinny, my chest needs development...Everyone says I look old for my age. I think I'm pretty tall now...I hope I'm not short at seventeen. I hope I'm big and won't shrink.

Compared to his friends he is "the small one." Joey frequently cites his short stature as a prominent defect, wishing soon that this would change. Whether describing basketball or street fights, he usually refers to his size. Though generally he degrades himself for being short, sometimes the emphasis is on being "tough anyway," or on "not taking it," no matter what the opponent's size. There is something very attractive for Joey about his "Jewish friends'" seeming lack of concern for these dimensions. In addition, even if he is physically lacking, it is his cleverness, the mental qualitities he can assert. The Jews represent in part for Joey a group in which his physical handicaps are not emphasized. In this respect they help him to mask blemishes. The intelligence and education that he credits them as symbolizing are, however, second best; were Joey strong and as physically able as he desired, the "brain" qualities would not be quite as attractive.

Joey does not even fully qualify for membership in the "Jewish" group, for his image of his intellect is also not an exalted one: "I don't like to look like a doctor, they're supposed to be real smart and know a lot. I'm not." Until his senior year of high school, Joey felt he could "get by," but he was not going to be very successful academically. "Almost making honors" his last year at school was a genuine surprise. He had studied more assiduously than ever before and most unexpected of all was the suggestion that college would be a feasible choice for him. To the observer, Joey's image of himself as "not bright" is less evident than his condemnation of his physical stature. To Joey his outstanding defect is size and puniness; after that, almost as if linked, come his intellectual failings.

What is important in this self-depreciation is that none of these estimates is absolute. On the two standards against which Joey measures himself he received a low rating. However, he can *always improve*. By going to the "Y" for physical development, as he did, he can build up his muscles. By standing up for his rights "on the block" he can counter any impressions of weakness. Intellectually, he continues to demonstrate improvement to himself. The groups he joins are among the "brains" of the school. His grades have risen to the point at which he now has "about the best report card in my home room." He believes in and is committed to *personal progress*.

There are more stable and consistently positive features in Joey's self-image. Reliability, steadiness, honesty, and service are all qualities he admires in several adult figures and tries always to emulate. He believes himself "popular," but never popular enough to be content.

The equation is almost quantitative; the greater the number of friends I have—the better I am as a person. Finally, there are the associated qualities of "independence" and dignity to which Joey clings despite his longings for popularity and his often degraded self-image. To a large extent this is in reaction to many dependent trends. Even when his dependecy is taken into account, however, Joey's insistence on dignity is not fully explained. In his dealings with employers, friends, and their parents, Joey indicates in no uncertain terms where his limits of compromise are. To yield beyond these lines—whether to please others or achieve further acceptance—is forbidden. The sources of this important theme, pride and independence, are discernible in models and heroes in Joey's life. A major source of support for the enhancing self-esteem rests with Joey's mileu as well. The responses of others are uniformly encouraging: promotions, the change from basement to "the floor"; an increasing number of friends from the prestigious Jewish group; "almost making honors"; followed by being told to "apply to college."

CHAPTER 7

The Roots of Identity Foreclosure: Three Theoretical Perspectives

INTRODUCTION

It is obviously important to bear in mind that these are findings derived from a *small, matched interracial sample*. They are not results that are in any way representative or "typical" of a "black ghetto" or of a "white slum". Such a leap in reasoning is ever tempting and fallacious. Nonetheless, even with this note of caution, we are left with a serious problem: How are we to understand findings which show that between 1962 and 1967 these black adolescents had different identity formation patterns than a sample of white adolescents? And the divergence in this development line is not a subtle one. Indeed, using several forms of comparison, the blacks' identity development appears to differ from that of the whites.

There are three perspectives we can take toward comprehending the black identity foreclosure. The perspectives do not contradict one another. Rather, they have different understandings to offer; they are different ways into the same phenomena. To be sure, investigation from each of these perspectives would require several separate studies. All three approaches will be sketched here so as to delineate what is required for a full understanding of these complex findings. Clearly, no one perspective is itself sufficient to explain the apparent variations in identity formation.

THE CONTEXT OF IDENTITY FORECLOSURE

An essential aspect of identity foreclosure is stasis; the individual's self-images show little to no change as he grows older. Inherent in these static self-images is the individual's experience—conscious and nonconscious—of the restricted alternatives. Sociocultural and intrapsychic forces underlie this experience. Proceeding first from the individual's sociocultural context, there is the *actual* fact of limited choice. This fact is referred to again and again by Frankie and the other black subjects in speaking of racial restrictions placed on part-time jobs, parties and dates, recreation opportunities and housing. They notice it most directly in the area of work opportunities. There were generally few part-time and summer jobs available; and those available positions seemed always associated with distasteful kinds of work. White adolescents found jobs that were, temporarily at least, satisfactory; and within the job there was the opportunity to "progress."

It is not only in the present that work is both limited and unsatisfactory for the blacks. Prospects for a broader range of jobs after graduating from high school seemed just as unlikely to them. The armed services was one of the few systems in which the black adolescents expected to find some chance of recognition and even—perhaps—advancement. Except for the military, the blacks envisioned a future of dull distasteful work. At best, there was the possibility of some day being a "boss" of a small store.[1]

A third set of environmental constraints involved heroes, those admired figures so important in adolescence. For the blacks, there were few men who were idealized in any way, few adult figures considered worthy of emulation. Most of these subjects flatly stated that they had no heroes, no people they wished to resemble either now or in the future. Occasionally, popular blacks such as Joe Louis, Jackie Robinson, Sidney Poitier, and James Baldwin were admired. None of the subjects considered the potent talents of these men as being within their reach. In the preceding chapter, we observed that the heroes were not only few in number but their number gradually diminished over time. Both the size of the list and its continuous diminution stood in marked contrast to the whites' large, ever-expanding list.

Related to both the limitations of work and heroes is the fantasied future. Take a job market that is small and unattractive. Then add few—if any—attractive or appealing adult examples of how a boy might appear in 10 years. Then, as if these facts did not make a dismal enough picture, add restrictions as to where the black man may live and play. To further stigmatize the future there are an abundant number of men who are on the street corners: the unemployed, the drunks, the "bums"; these are men who provide the black adolescents with what Benny called "the Negro image."[2] Such models of the future are seen

---

[1]Not only do the subjects in the study observe these *actual* work limitations; Pierce (1968), Grier and Cobb (1968) and Clark (1965) make similar observations.

[2]See Benny's description of this notion on p. 68.

each day. Even if they are not seen "on the avenue," "Moms" is around to remind her son of his likely fate. In the instance in which his mother is not forecasting such unhappy destinies, frequently the black youth learns of his or his close friend's father who has deserted, or who remains and is the object of constant disdain, such as Frankie's "Pops."

From multiple sources the message is clear. For the working class black adolescent there are few options. Still another message is given by these constraints, that of devaluation by the community. Erikson speaks of the importance to the adolescent that he be "recognized" by the community around him:

> ... we speak of the community's response to the young individual's need to be "recognized" by those around him, we mean something beyond a mere recognition of achievement, for it is of great relevance ... that he be responded to and given function and status as a person whose gradual growth and transformation make sense to those who begin to make sense to him.[3]

Opportunities for new functions and status by way of work are sparse in both present and future. This impoverishment can only lead to the conclusion by the "young individual" of the community's misrecognition—or nonrecognition—of him. Restrictions in living area, recreation, and heroes further inform the black adolescent of his inferior standing in the community. The thrust of these degrading experiences is toward self-limitation and ever-diminishing self-esteem, two themes frequently touched on by the adolescents in their twice yearly interviews.[4]

Confirmation of this inferior status is given as well by his personal history and of the history he knows of his race. Each black in the study made it clear that he wished not to speak of his own past,[5] while knowing more than glimmerings of the broader past of slavery, submission, and ever repeated degradation. Grier and Gobb describe this evaded knowledge of the past:

> History is forgotten. There is little record of the first Africans brought to this country. They were stripped of everything. A

[3]Erikson (1968), p. 156.

[4]The evidence supporting "low self-esteem" among blacks is highly problematic as several recent reviews have pointed out. (This is discussed in Chapters 8 and 9). At the level of our *interview data*, the diminished self-esteem was clearly present for the black subjects. In our next chapter, we discuss the recent data and reviews regarding the question of black self-esteem.

[5]The Q-sort results of high negative correlations of black boys' present with their past self-image is consistent with this observation.

calculated cruelty was begun, designed to crush their spirit. After they settled in the white man's land the malice continued. When slavery ended and large scale physical abuse was discontinued, it was supplanted by different but equally damaging abuse. The cruelty continued unabated in thoughts, feelings, intimidations, and occasional lynchings. Black people were consigned a place outside the human family and the whip of the plantation was replaced by the boundaries of the ghetto.[6]

A most difficult conceptual question is posed by the interview data. It concerns the complex interface between person and context: How do the historical and contemporary forces noted above influence personality development in black adolescents? Why, for example, does our sample show a pattern of identity foreclosure while other blacks have apparently been free of this impairment in identity development? Several sociological and sociopsychological studies suggest certain variables which likely mediate between the black adolescent and his context.

Numerous analyses of black Americans discuss devalued, constricted life styles together with some of the variables underlying their persistence (Clark, 1965; Billingsley, 1968; Elkins, 1963; Pettigrew, 1964). In terms of social class, Billingsley notes that there are several strata of lower socioeconomic class blacks, representing 50% of all American blacks. The black subjects in this study were members of the two predominant levels of this class—the working poor and the nonworking poor. There are several limitations imposed upon these two strata:

... to be living poor and black in 1968 means severe restrictions in the most basic conditions, particularly focused in the areas of family income, education of parents, occupations of family heads, family housing, and health care.[7]

In his extensive review, Pettigrew (1964) summarizes accumulated data concerning psychological consequences resulting from this powerful combination of racial and economic oppression. In particular, there is the impact accompanying the "subordinate social role" into which the lower class black has been placed:

---

[6]Grier and Cobb (1968), pp. 25-26. Since the original writing of this chapter, the heightened interest in black history and "black studies" has obviously modified this stance.
[7]Billingsley (1968), p. 142.

The effects of playing this Negro role are profound and lasting. Evaluating himself by the way others react to him, the Negro may grow into the servile role; in time the person and the role become indistinguishable. The personality consequences of this situation can be devastating—confusion of self-identity, lowered self-esteem, perception of the world as a hostile place, and serious sex role conflicts.[8]

Yet, as both Billingsley and Pettigrew recognize, not all American blacks are crippled by these social conditions. To understand the non-foreclosed blacks, the "men of distinction" (Billingsley) means coming to terms with those factors which permit mobility (psychological and social) within a context of limited opportunity. Billingsley (1968) analyzes the history of several socially successful blacks dealing specifically with their sources of social support. These supports ranged from a familial history of special status during slavery, to money and prosperity. Prominent within the lives of the men he analyzes were special role models or personal "heroes":

> Most important ... were the Negro role models available to him who took a special interest in him. Many a Negro boy has been encouraged in his social achievement by the kind and sometimes forceful hand of members of the black establishment. Thus the Negro minister, school teacher, doctor and occasionally undertaker, provided a kind of support often absent in urban ghetto today.[9]

Such models were consistently omitted by the black subjects in the many interviews over the four years. As noted earlier in this section, the blacks were most aware of their "anti-heroes." The seeming lack of fluidity in these black adolescents may well be one of the consequences of relatively unavailable role models.

To be sure, the conditions fostering identity foreclosure cannot be dealt with in a single factor explanation. As the interviews of our subjects, and the above formulations suggest, behind the black identity foreclosure is an intersection of vastly limited social resources and historical as well as current devalued stereotypes.[10] At the level of the individual black, an impoverishment of role models may well be the clearest final result of this complex range of constricting forces. Figure 1 summarizes the preceding discussion of the psychosocial matrix of black identity foreclosure.

[8]Pettigrew (1964), p. 25; Rainwater (1966) also discuss the consequences of devalued roles for lower socioeconomic class blacks. His discussion treats at some length the issue of self-stereotyping ("victimization").

[9]Billingsley, 1968, p. 142.

[10]Elkins (1963) deals with historical roots to these stereotypes, while, as noted previously, Pettigrew (1964) and Rainwater (1966) particularly discuss contemporary aspects of the stereotyping.

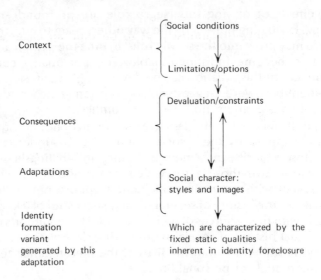

Context
{
Social conditions

Limitations/options

Consequences
{
Devaluation/constraints

Adaptations

Social character:
styles and images

Identity
formation
variant
generated by this
adaptation

Which are characterized by the
fixed static qualities
inherent in identity foreclosure

Figure 1. The context that evolves and maintains identity foreclosure: a summary.

THE DEVELOPMENTAL PERSPECTIVE

We have thus far been concentrating upon the context in which identity foreclosure emerges. If we now ask how that context has impeded or enhanced growth of the individual, we shift to a developmental approach. What in the black adolescent's own history might underly his developing in the direction of identity foreclosure?

Erikson's epignetic model of development is most pertinent to these questions (Erikson, 1950; 1968). According to this developmental theory, there are two points in a person's history where disruptions could lay the groundwork for later emergence of a forclosed identity.

One of the historical points is in the latency period. Within this developmental phase, directly preceding adolescence, "competence" is a prominent concern for the developing child:

... he must begin to be somewhat of a worker and potential provider before becoming a biological parent. With the oncoming latency period, the normally advanced child forgets, or rather "sublimates" (that is, applies to more useful pursuits and approved goals) the necessity of "making" people by direct attack or the desire to become mother or father in a hurry: he now learns to win recognition by *producing things*. He develops industry ... to become an eager and self-absorbed unit of a productive situation to completion is an aim which gradually supersedes the whims and wishes of his idiosyncratic drives and personal disappointments ... He now wants to make things well ... He develops the pleasure of work completion by steady attention and persevering diligence.[11]

Two general conditions are required for the development of this competence: previously successful steps in earlier ego development and a facilitating environment within the latency period. If these requirements are fulfilled, the set of conflicts—or "crisis"—of latency then resolves and the developing child emerges with "a sense of industry." The identity precursors of such a successful resolution are positive "work identifications."[12] The successful resolution is represented in the next stage of ego development—adolescence—in the sense of "anticipation of achievement."[13]

However, if the individual fails to resolve these latency issues, he emerges from this period with a sense of "inadequacy and inferiority." The identity precursor corresponding to this developmental failure is *identity foreclosure*, a premature interruption in the adolescent task of identity formation. There are two seemingly opposite forms in which this precursor can be expressed. It can appear *negatively*, as the "sense of inferiority, the feeling that one will *never* be any good,"[14] or it can appear *positively*, when, for example:

[the child] identifying too strenuously with a too virtuous teacher or becoming the teacher's pet ... his sense of identity can become prematurely *fixed* on being nothing by a good little worker or a good little helper, which may not be all he *could* be.[15]

[11]E. Erikson, "Growth and Crises of the 'Healthy' Personality," p. 86.

[12]E. Erikson, "Ego Identity and the Psychosocial Moratorium," in H. Witmer (Ed.), *New Perspectives for Research on Juvenile Delinquency*, Washington, D.C.: U.S. Government Printing Office, 1955, p. 8.

[13]Ibid.

[14]E. Erikson, "Growth and Crises of 'Healthy' Personality," p. 87; (italics added). It is of interest that Erikson adds that this is "a problem which calls for the type of teacher who knows how to emphasize what a child *can* do and who knows a psychiatric problem when he sees one."

[15]Ibid., p. 88; italics of "fixed" are added.

The predominant form of identity foreclosure for the black subjects was a negative one. Repeatedly in the interviews, as illustrated earlier, there emerged the themes of inferiority, mediocrity, and degradation. Sufficient data is not available to firmly reconstruct the latency of these subjects. Their accounts of work situations as adolescents certainly encourages the speculation that experiences of failure at work, be it with school or "jobs," earlier in their lives as well. The black youth complained with great frequency of "being looked down upon" in school and of being given second-rate work opportunities, "where doors are slammed at you." Events such as these in latency would be compelling conditions for the generation of a sense of inferiority and the corresponding identity precursor, identity foreclosure.

There are intrapsychic as well as other sociocultural determinants that may underlie the development of a "sense of inadequacy and inferiority" with its concomitant identity foreclosure:

[The sense of inadequacy] may be caused by an insufficient solution of the preceding conflict: he may still want his mummy more than knowledge: he may still rather be the baby at home that the big child in school; he still compares himself with his father and the comparison arouses a sense of guilt as well as a sense of anatomical inferiority. Family life (small family) may not have prepared him for school life, or school life may fail to sustain the promises of earlier stages in that nothing he has learned to do well already seems to count one bit with the teacher. And then again, he may be potentially able to excel in ways which are dormant and which, if not evoked now, may develop late or never.[16]

One type of identity foreclosure is a derivative of the stage of development preceding latency. This foreclosure is one of *total negation*, of repudiation. It is based on "all those identifications and roles which, at critical stages of development, had presented to the individual as most undesirable or dangerous. . . ."[17] This foreclosure is that of *negative identity*.[18] The individual's identity configuration is prematurely fixed on the repudiated, the personally scorned and rejected identifications and roles.

[16]Ibid., p. 86.

[17]E. Erikson, "The Problems of Ego Identity," p. 131.

[18]In Erikson's writings there is interesting support for viewing negative identity as a subtype of identity foreclosure. Instead of using the term "negative identity" in speaking of the derivative for adolescence of a developmental failure in the oedipal period, now the term "*role fixation*" is employed. See "The Problems of Ego Identity," p. 144 and Identity: Youth and Crisis, p. 94.

Negative identity is the adolescent derivative of a failure to resolve the set of issues belonging to the developmental stage of conscience formation. It is this phase of ego development in which the conflicts between guilt and initiative become most prominent. Where guilt predominates, there results " a *self-restriction* which keeps an individual from living up to his inner capacities or to the powers of his imagination and feeling."[19] This pattern of self-restriction is already familiar as a fundamental aspect of identity foreclosure. However, where conflicts over initiative and gult are of great magnitude, there may emerge *paralysis* of ambition and a choice of the despised as the focus of one's self-image. This configuration, then, is negative identity:

> Such vindicative choices of a negative identity represent, of course, a desperate attempt at regaining some mastery in a situation in which the available positive identity elements cancel each other out. The history of such a choice reveals a set of conditions in which it is easier for the patient to derive a sense of identity out of a total identification with that which he is at least supposed to be than to struggle for a feeling of reality in acceptable roles which are unobtainable with his inner means ... the relief following the total choice of a negative identity.[20]

One expression of negative identity, of relevance here, is that of "snobbism." Erikson discusses forms of snobbism as a response to the danger of identity diffusion. Confronted with the adolescent conflicts of identity formation and the possibility of failing to resolve them, an earlier developmental failure becomes most prominent:

> ... they [upper-class forms of snobbism] permit some people to deny their identity confusion through recourse to something they did not earn themselves, such as their parents' wealth, background, or fame or to some things they did not create ... But there is a *"lower-lower"* snobbism too, which is based on a pride of having achieved a semblance of nothingness.[21]

Negative identity is a highly specific type of identity foreclosure. Identity formation is prematurely halted through a commitment of the

---

[19]Erikson (1986), p. 120; italics added.
[20]Ibid., p. 176.
[21]Ibid., p. 176; italics added.

individual to that which is alien, personally most shunned. In making so *total* a choice, the ambiguity and conflicts inherent in multiple identifications are eliminated in favor of a single principle, a single set of identifications. The identity rests on the despised. The individual considers himself as hateful, *totally* undesirable.

In another context Erikson refers to this identity pattern as a "total commitment to role fixation."[22] Although this fixation may become most obvious in adolescence, its genesis is in the oedipal period of life. Failure to deal with and, on balance, to overcome the conflicts over guilt and initiative have facilitated the emergence of a negative identity in adolescence:

> ... the display [negative identity] ... has an obvious connection with earlier conflicts between free initiative and oedipal guilt in infantile reality, fantasy, and play ... the choice of a self-defeating role often remains the only acceptable form of initiative ... this in the form of a complete denial of ambition as the only possible way of totally avoiding guilt.[23]

Negative identity represents an identity configuration that is an integration of hated and repudiated identifications. It is thus a configuration of self-repudiation, or self-negation. Along a continuum of identity foreclosures, negative identity represents the negative pole. Between this polar type and the positive polar type—the "precocious genius" or "pet"—lie the many other variations of identity foreclosure. Within these poles are varieties of "inferiority" and "superiority." The interviews suggest that the black subjects' identity foreclosure falls in the negative range of such a continuum. They consider themselves "inferior," limited, "lame," and therefore fit *only* for a certain niche.

Yet within this group of adolescents the extreme polar types, negative identities, may also exist. Careful scrutiny of early personal history, early identifications, and of the content of a subject's multiple self-images *as well as* their structural relationships would be required to discover the presence of negative identity configurations. Are the current self-images of a subject, for instance, consistent with those roles and figures he was taught were most undesirable? Do the subject's highly valued self-image features correspond to those features of others that were regarded in early years as despicable and degraded?

---

[22]Ibid., p. 184.
[23]Ibid.

Disturbances in two different developmental periods may underlie identity foreclosure patterns. Clearly, the roots of these disturbances are multiple. For the black subjects it is likely that issues of work in latency play a prominent role in the identity foreclosure. However, a number of other conditions have also suggested themselves as significant in the genesis of this most striking variant, among them being the important figures who are internalized by the developing child: fathers, the heroes, and the anti-heroes. In conceptualizing negative identity as a type of foreclosure, these identifications must be carefully considered as a realm of determinants.

The developmental perspective points to late childhood and latency as critical phases associated with identity foreclosure variants. Within either of these phases, failure to resolve crucial sets of issues will lay the groundwork for the later emergence of identity foreclosure or negative identity. Figure 2 outlines the dimensions and relationships highlighted by the developmental perspective.

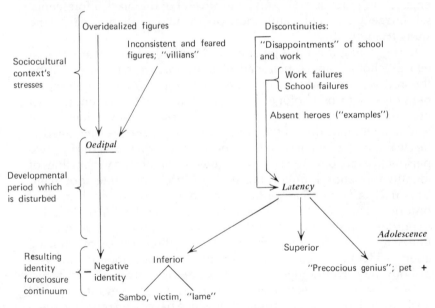

Figure 2. Developmental model of identity foreclosure.

THE FUNCTIONAL PERSPECTIVE

A third approach in analyzing identity foreclosure takes account of relationships with ego functions. An important group of ego functions are those which involve cognition. Our general hypothesis about cognition and identity formation states that cognitive processes are a prominent set of functions underlying patterns of identity formation. The influence of the cognitive processes is mediated through their interactions with self-images and self-image systems. The details of how cognitive functions interact with self-images remain to be empirically determined. At issue here, for example, is the problem of whether some cognitive functions constrain change within self-images, while others facilitate change in the individual's system of these images and thus in his mode of identity formation.

The results of our more microscopic self-image analyses studying polarization and complexity within the black and white adolescents' self-images suggest that cognitive/perceptual processes associated with identity formation may also differ between the two groups of adolescents. In "structural complexity" analyses, we looked at the arrangements of self-statements *within* each self-image. The results of these further analyses support the speculation that blacks may be engaged in more self-stereotyping. Several of the black adolescents' self-images were significantly more polarized over all four years of the study (Hauser, 1972).

In suggesting this third perspective to identity foreclosure, that of ego psychology and functional relationships, we have stressed the theoretical model which could be applicable. For a better comprehension of ego psychological aspects of identity formation, this last perspective and the studies it entails are crucial. However, it must not lead to a minimization of the larger context and the pressing need to be clear about its impacts on these personality processes. The first two perspectives are directed toward this environment. The variant lines of identity formation found in the study of black and white adolescent boys may in large measure be a consequence of sociocultural and historical forces. On the basis of both scientific and moral reasons, it is important that these forces and their impacts upon the individual be understood; for to make changes in the direction of minimizing abortive identity development will require rigorous empirical analyses of both the surroundings and the developing individual.

Our next two chapters review and discuss those relevant empirical analyses and theoretical formulations which have been reported since the original publication of these results.

CHAPTER 8

Empirical Studies

This chapter deals with empirical studies published over the last ten years which have investigated racial differences in terms of several personality and social psychological dimensions. We have selected those dimensions and investigations which are most relevant to understanding identity formation and adolescent development. These many studies are presented through our initial summary comments, and then more extensively through a detailed table.

Our text summary, covering five areas, is intended to draw together findings which are most germane to the original study, presented in the preceding chapters.[1] The review table included in the chapter has a broader purpose. Here we give the reader a more detailed picture of much of this recent research and thereby allow for closer scrutiny of important differences among the investigations, such as aims, sample characteristics, instruments, and design. In our final section we discuss several factors which must be considered in order to grasp the basis of the contradictory and often confusing findings reported in the studies.

In terms of the inclusion and exclusion of material, the reader will note some degree of overlap in this chapter. Findings have been included in the text summary but not the table, when they have been found to be relevant but do not fit the format of the table (for example, case studies). There are also a number of studies included in the table but not the summary; these are of two types. One type involved studies whose findings are more methodologically-concerned and of interest in terms of empirical approaches. The other group of studies not directly related to the issues discussed in this chapter, is included because of relevance to the theoretical material presented in Chapter Nine.

---

[1]The purpose is not a general evaluative investigation of the work in the field (for example, see Brand, Ruiz and Padilla, 1974, in Table); rather, we examine those findings which tend to support or contradict the original findings.

This review is offered not as an exhaustive description of the work in this area, but as a well-rounded and extensive sample of empirical investigations.[2] In most cases, where many conclusions were offered by the authors, only those most relevant to our work were included. In other cases, additional pertinent information was found to be too lengthy and detailed to summarize within the brief format of the table. For these studies the interested reader may wish to refer to the original source.

One last introductory comment concerns the somewhat artificial distinction made here between the empirical and theoretical writings. Obviously, empirical studies are based at some level on theoretical formulations; and theoretical arguments cite empirical work to support their positions. We present the empirical studies prior to the more theoretical work in order to most clearly highlight certain issues, and summarize the many findings, which are then discussed at greater length in our final chapter.

SELF-ESTEEM[3]

Numerous studies of black and white differences in self-esteem dimensions have been reported since 1971. Many were reviewed by Rosenberg (1979) and Wylie (1978) in their recent monographs. Our original study did not deal with self-esteem through direct measures. However, much of the later discussion and interpretation of the contrasting findings included observations about diminished self-esteem for the black adolescents. In now turning to the vast number of systematic self-esteem studies, we are presented with findings that strongly argue against any general conclusions that black adolescents have lower self-esteem. Although one can certainly argue that these paper-and-pencil instruments are not assessing to the same self-evaluative dimensions as were our clinical interviews, it is still important to recognize that the many studies are generating conflicting results. Consequently, to generalize from our limited sample to broad assertions about black (or white) self-esteem is problematic. Besides the small sample limitations presented by our sample, there is the

[2]This is, in general, an unbiased review. Studies have been included which do contain methodological flaws, such as confounding of race and class variables; (for example, see Davids, 1973, Miller, 1975, Gestinger, et al. 1972). For a more evaluative review, please see Wylie, 1979.

[3]For the moment, we will be using the term "self-esteem" to cover many similar terms, as seems to be a general practice in the literature. Later, we discuss the problems raised by the proliferation and near-equivalent usage of terms such as self-esteem and self-concept.

additional difficulty created by the fact that our "self-esteem" observations were based on extensive longitudinal interviews rather than the self-report test data used by most other studies in the literature. In what follows, a few illustrative examples convey three types of findings: studies which show blacks to have lower self-esteem than whites; those which report black subjects as having higher self-esteem; and those which find essentially no difference between blacks and whites.

(1) Blacks have lower self-esteem than whites:

Crain and Weisman (1972) found black adults to have, on the average, lower self-esteem than whites. This was partly due to the larger spread in self-esteem scores of blacks (the scores ranging from very high to very low). When geographic and racial variables were examined separately, Southern blacks were found to have lower self-esteem than whites, and Northern blacks higher self-esteem than whites. This difference was explained in terms of the increased opportunity for expressing aggression in the North. Brown (1977) described the self-perception of poor black adolescents as depreciated, incompetent, and ineffective.

We do not mean to suggest that there are only two studies which have found blacks to have lower self-esteem. However, in our review of the recent literature, we found an overwhelming proportion of findings showing blacks to have equal or higher self-esteem than that of whites. This is striking when one considers how strongly supported the notion is that blacks have lower self-esteem, by theoretical, political, social, even common-sense reasoning—indeed, in every way but empirically. Banks (1976) suggested that we have the situation of "a paradigm in search of a phenomenon." We will return to this important issue in the following theoretical chapter.

(2) Blacks have higher self-esteem than whites:

Reece (1974) summarized four independent research studies whose results indicated that many black children and youths hold more positive self-images, than those results from studies published ten or twenty years ago, when much data indicated that those who were disadvantaged and/or black tended to have lower self-esteem. Trowbridge and colleagues (1972) also report that working-class children and adolescents have higher self-esteem than comparable white subjects. The findings of Powell and Fuller (1972), and more recently, Cicirelli (1977) support the pattern of higher self-esteem for blacks. Cicirelli also found an effect of welfare status. When the two variables of race and socioeconomic status were looked at separately, the self-concept of non-welfare black children was no higher than that of the white group. The differences, then, between the blacks and whites, were considered to be mainly a function of the welfare status of the subjects.

(3) There is little or no difference in self-esteem between blacks and whites:

A number of studies describe no self-esteem difference between black and white subjects. Zirkel and Moses (1971), Rosenberg and Simmons (1972), Heiss and Owens (1972), Davids (1973), Healy and DeBlassie (1974) and Busk, Ford, and Shulman (1973), found that over-all the self-esteem of blacks and whites did not differ significantly. These results are consistent with Yancey, Rigsby, and McCarthy (1972) who reported no evidence of a systematic effect of race (on self-esteem or on psychological symptoms of stress) when other relevant dimensions of social position were taken into account. A major point stressed by a number of these studies is the importance of controlling for social class. When this dimension is held constant, many seemingly racially-based self-esteem differences are no longer found.[4]

RACIAL PERCEPTIONS AND PREFERENCES

Studies of racial perceptions, trait attribution, stereotyping, and discrimination are relevant to identity development because they address individual perceptions of the social reality (how the person views the surrounding society and his or her place in it, based on whom one is like or unlike).

In 1939, Clark and Clark began publishing work that has become classic in the area of racial identification and preference, in terms of its social and psychological ramifications as well as the number of times it has been cited and replicated. Their basic design (Clark and Clark, 1947) involved using black and white dolls, presenting young black children with the following requests:

1. Give me the doll that you want to play with.
2. Give me the doll that is a nice doll.
3. Give me the doll that looks bad.
4. Give me the doll that is a nice color.
5. Give me the doll that looks like a white child.
6. Give me the doll that looks like a colored child.
7. Give me the doll that looks like a Negro child.
8. Give me the doll that looks like you.

Their most important finding was the black children's preference for the white doll, this being interpreted by the Clarks as a rejection of

[4]Even when sociocultural variables are examined, some researchers (for example, Gestinger, et al., 1972) have concluded that race and/or SES are not significantly related to self-concept—or that the measures employed are not sensitive enough.

their blackness and a desire to be white. For the most part, recent applications of this approach indicate that favorable racial preferences and attributions are now made by black subjects.

In an updated and modified version of Clark and Clark's Dolls experiment, Hraba and Grant (1970) observed that the majority of children preferred dolls of their own race, regardless of color. Teplin (1976) in a photo-choice type variation of the dolls procedure also found that the majority of white children chose pictures depicting whites; and black children who made more black preference choices also had higher self-esteem. The findings of Stabler and colleagues (1972) conflict with these studies, (though actual methodological similarity may be questioned), as they found that children associated negative self-statements to the color white. The authors suggest that attitudes toward the colors black and white may influence the way black and white children view each other and themselves. An underlying assumption in these studies is that doll (color) preference generalizes to a preference towards people. This leads us into the studies which look at trait attribution and racial perceptions.

Brigham (1973) related blacks' and whites' perceptions of a group's attitudes to the actual expressed attitudes of the particular group, and found the degree of the relationship between them to vary according to the sample. In a later work, also looking at trait-stereotyping, Brigham (1974) found within-race agreement to increase with age, and the largest amount of between-race agreement between grades 7-11. Both samples (black and white) showed a significant tendency to attribute favorable traits to their own race, this tendency being significantly greater within the black sample. Similarly, Terry and Evans (1972) found both black and white respondents to attribute more discrimination to other races and classes than to themselves. They comment that the variable of social class may be a more important factor than race in affecting discrimination behavior.

As the above findings illustrate, members of a minority group do differentiate in the perception of personality characteristics of their own group and of the majority group (Bayton, Austin, and Burke, 1972). Along the same lines, Polite, Cochrane and Silverman (1974) found that more positive self-perception in black college youth was directly related to the degree of perceived differences between the blacks (their own ethnic group) and other ethnic groups. And in an extension of the positive racial perceptions theme, Teahan and Podany (1974) comment that an increase in pride about race (for blacks) may result in more negative or critical attitudes toward whites.

GROUP MEMBERSHIP

The findings in the third topical area concern issues of integration and segregation, and minority versus majority status of racial groups. These studies are closely related to the investigations of racial percep- tions. Regarding the impact of group membership, Busk and col- leagues (1973) found that the most influential factor affecting adolescent students' self-concept was not ethnic membership, but the racial composition of school attended. More specifically, Powell and Fuller (1972) report that black adolescents in segregated schools have higher self-esteem than blacks in desegregated schools.

LOCUS OF CONTROL

Another dimension thought by some to be related to self-concept is "locus of control," the degree of mastery the individual believes he has over events in the external world. Milgram (1971) found no differ- ence in internal locus of control between races. Crain and Weisman (1972) reported that the individual's sense of internal control was inde- pendent of self-esteem, at least for blacks. Garcia and Levenson (1975), on the other hand, did find racial and class differences on a locus of control measure, these varying according to specific subscales (see Table below). In a more recent work, Taylor and Walsh (1979) examined the effects of race and occupation on "system-blaming" and its possible function in the preservation of self-esteem. They found that blacks and those in higher status occupations were more likely to blame the system than whites and those in lower-status occupations. They suggest a positive effect of system-blaming on self- esteem, the idea being that "attributing failures to the social system serves as an alternative to internalizing responsibility and suffering diminished self-esteem" (p. 248).

FATHER ABSENCE AND ROLE MODELS

Our final set of studies analyzes the effects of father absence and the more general question of availability of role models. These are themes directly touched on in the *Black and White Identity Formation* inter- views and interpretations.

Father absence was found to have different consequences by race (Hunt and Hunt, 1975). The authors report some achievement and identity "costs" for white boys, and the possibility of some positive identity "gains" for black boys. This pattern was consistent across the social class levels. Hartnagel (1970) observed that black father-absent boys had significantly less discrepancy between actual and normative self-potency than white father-absent boys. "Normative self" refers to attitudes of others with respect to what one should be, and "actual self" refers to attitudes of others with respect to how well one measures up to these normative expectations. In a more general study (no racial focus or comparisons), D'Andrade (1973) found more feminine/less masculine self-description characteristics in the sex-role identifications of all males from father-absent homes. Earl and Lohman (1978) offer data with respect to actual behaviors in the area of father-absence and role models. In their study of black boys and mothers from father-absent families, they found that in half of the cases, the boys did see their fathers with reasonable frequency. Furthermore, all the boys were found to "have access to some black male who could, at least potentially, serve as a role model" (p. 414).

As the reader must now be well aware, our brief survey of the empirical research conducted over the last ten years leaves us with conflicting and somewhat confusing results. How are we to make sense of this situation?

One possibility is through a methodological critique. The table has been provided to allow the possibility of more closely inspecting the samples and techniques. Brand and coleagues (1974) have reported such a review. A similar critique is presented by Gordon (1976) who analyzed eighty-three studies in terms of methods, hypotheses, geographical location, and other relevant variables. This investigation led her to conclude that either "researchers have been studying a widely shifting phenomenon or that there are serious validity problems in the study of self-concept" (p. 381). Gordon further suggests that "the study results, taken as a whole, appear to be largely a function of the research instrument used; the theoretical orientation of the researcher, and the time and place of the study" (p. 381). Most recently, Wylie has extensively reviewed and commented upon racial self-esteem studies (Wylie, 1979).

In addition to the methodological differences, there are at least two basic conceptual distinctions to be considered. The first involves a confusion of terms. The literature provides many different self-describing terms: identity, self-esteem, self-concept, self-identification, self-image, self-evaluation. There are at least two ways in which an error in understanding may thereby occur: either different terms are being used to describe the same entity or process; or, similar terms are being mistakenly interchanged to discuss different phenomena. For an example of this conceptual vagueness, and related instrument issues, consider the following: Zirkel and Moses (1971) and Trowbridge, et al. (1972) used the Coopersmith Self-Esteem Inventory to study "self-concept," while Miller (1975) used it to study "self-concept" and "self-esteem." On the other hand, Ward and Braun (1972) used the Piers-Harris Self-Concept Test, and Powell and Fuller

(1972) used the Tennessee Self-Concept Scale—both to study self-esteem.

These are just a few examples of the rampant interchangeability of terms found in the literature. We are left with many questions as to what *is* being studied, what the tests do, in fact, measure; and how, if possible, results could be compared.

The other important conceptual distinction to be considered is that of psychological perspective. As the reader will see from our discussion in the next chapter, a researcher's particular orientation is an integral aspect of the entire research endeavor. Attempts to compare studies "across the board" will prove to be misleading, if these perspective differences are not taken into account.

EMPIRICAL STUDIES RELEVANT TO RACIAL DIFFERENCES IN IDENTITY DEVELOPMENT[A]
(Self-Esteem, Self-Concept, Identity)

Study	Purpose	Description of Subjects	Instruments, Methods	Major Findings, Author's Conclusions
August and Felker (1977)	To ascertain whether the empirical relationship between self-concept and affective learning style is applicable for black children, as well as middle- and lower-class children of both ethnic groups.	80 5th-grade children from 4 elementary schools of varying racial compositions, in Northwestern Indiana.	Piers-Harris Self-Concept Scale.	Self-concept was found to interact significantly with race. "...The high-self-concept white children recalled positively rated words more readily than negatively rated words, while their low-self-concept peers manifested no memory predilection. Although the low-self-concept black children also reflected no preference for their affective evaluations, the high-self-concept blacks showed a greater propensity to recall their negatively rated words" (p. 253). Social class was found to have little influence on the affective learning style of the Ss.
Davids (1973)	To investigate the development of the self-concept and mother-concept in a group of young black children. Designed to obtain additional empirical evidence pertinent to issues of blacks' internalization	42 black and white, male and female children, aged 3-6, attending either a YMCA nursery school or a day-care center in Providence, Rhode Island.	Children's Apperception Test (Bellak, 1954). Story Completion Technique (Davids and Lawton, 1961). Picture Self-Rating. Adjective Check-List.	No significant differences were found between the self-concept scores obtained from black and white children. Regardless of whether the self-concept is assessed by a projective procedure or an objective procedure, self-concepts of the children in this sample essentially similar.

[A] The table entries are roughly divided into categories which are discussed in the text. Information in the column headed "Purpose" consists of the aims stated by the article's author(s). Similarly, "Findings" and "Conclusions" are those expressed by the author(s).

Study	Purpose	Description of Subjects	Instruments, Methods	Major Findings, Author's Conclusions
	of whites' negative attitudes leading to low or negative self-concept.			No statistically significant mean difference between races on mother-concept.
				Correlation found between self-concept and mother-concept.
				Author suggests that absence, of difference in levels of self-concept and mother-concept between the black and white children may be because the Ss were at an age where the most important influence and determinant of self-esteem, is the home environment. These children may be valued quite positively in the home and smaller community and have not yet had to deal with the significant and traumatic degrading experiences (if they even exist to this degree) to be encountered in the larger society.
				(Author notes that no attempt was made to obtain exact information concerning SES backgrounds, nor to control for possible social class factors.)

Edwards (1974)

To determine significance of race (black/white) as an independent variable.

Part of a longitudinal study of how adolescent boys adapt to different high school environments.

Over 1,800 boys in 4 Detroit high schools

School 1—inner-city all-black high school with high rate population turnover (47% per year).

School 2—inner-city, serving predominantly black population with population turnover of 20% per year.

School 3—all-white suburban school, with population turnover of 16% per year.

School 4—all-white school, with fairly stable population (turnover of 8% per year).

Data collected in 1968, 1969, 1970.

Duncan's socioeconomic status (Blau and Duncan, 1967; Reiss, 1961).

The matrices (U.S. Employment Service, 1971).

Self-Esteem (Bachman, Kahn, Mednick, Davidson and Johnston, 1967; Rosenberg, 1965).

Rotter's Internal-External Scale (Rotter, 1966).

The Marlowe-Crowne Social Desirability Scale (Crowne and Marlowe, 1964).

The Social Exploration Scale (Edwards, 1971; Kelly, 1969).

"There were no significant and meaningful differences between blacks and whites on self-esteem, internal control, social desirability, or social exploration ... Other factors must be at work to cause the numerous black-white differences reported for self-esteem, internal control, and social desirability" (p. 47).

Conclusion of findings suggest that race is not a relevant variable for personality research or theory.

Study	Purpose	Description of Subjects	Instruments, Methods	Major Findings, Author's Conclusions
Krate, Leventhal and Silverstein (1974)	To assess change in identity among urban low-income black youth, using process-oriented procedure.	25 male, 25 female black students chosen at random from a Northern New Jersey state college. Black students attending this college were from predominantly low-income urban areas.	27 of the original 28 items devised by Hall, Cross and Freedle (1972) were recast into first-person statements and typed on 3x5 cards. Subjects were asked to rate deck of 27 statements on a 4-point scale (most to least like me) in 4 time perspectives (now, 4 years ago, 2 years ago, future).	There were no sex differences. The data suggest a shift away from a perceived past Negro identity toward a perceived present and future black identity.
Miller (1975)	To investigate the effects of maternal age, education, and employment status on children's self-esteem.	61 black inner-city and 97 white suburban 8th-graders and their mothers, representing educational ethnic, and social differences, chosen from a random selection of schools within 6 school districts.	Coopersmith Self-Esteem Inventory. Variety of demographic data obtained from each mother.	Mother's level of education was found to be significant for the main effect of sex, having an effect on child's self-esteem for inner-city black male sample. Lower levels of self-esteem were found for the male children whose mothers had less than a high school education. Full-time employment of mother had a greater effect on self-esteem for inner-city sample than for suburban sample.

| Taylor and Walsh (1979) | To test a major proposition and 6 explanatory hypotheses concerning to role of discrimination and segregation on the self-esteem of blacks and its (possible) difference from that of whites. | Black and white workers at 3 occupation levels (low, medium, and relatively high).

238 park workers (111 black and 127 white)—manual strata/all male.

191 mailcarriers (43 black and 115 white)—clerical strata/2 black and 1 white female.

144 high school teachers (29 black and 115 white)—semiprofessional strata/16 black and 47 white females. | Interviews and Questionnaires, focussing on "system-blame", and "self-esteem" concepts. | No significant relationship between mother's age and child's self-concept.

"The differences realized in this study support the concept that environment influences self-concept" (p. 142).

"...Results support recent studies showing black self-esteem at least equal to that of whites" (p. 250). (Please refer to original article for more detailed description of results of the six specific hypotheses.) |

137

Study	Purpose	Description of Subjects	Instruments, Methods	Major Findings, Author's Conclusions
Yancey, Rigsby and McCarthy (1972)	To evaluate the relative effects of racial status on a measure of self-esteem and on a measure of psychological symptoms of stress.	1,179 respondents from residential areas having lower-, working-, and middle-class blacks and whites. 362 blacks and 350 whites in Nashville, Tennessee. 215 blacks and 252 whites in Philadelphia, PA. Whenever possible, head of household was interviewed; if unavailable, second adult.	22-item scale of psychomatic symptoms of stress (Langer, 1962). Self-esteem scale (Rosenberg, 1965).	*No* evidence of a systematic effect of race when other relevant dimensions of social position are taken into account. In Nashville (the south), blacks showed higher levels of self-esteem and lower rates of reported symptoms of stress than whites.

(Racial Perceptions & Preferences/Trait Attribution—including Self-Attribution)

Study	Purpose	Description of Subjects	Instruments, Methods	Major Findings, Author's Conclusions
Bayton, Austin and Burke (1972)	To investigate perceptions of racial characteristics in the dimensions of sex and race.	240 black students (120 female, 120 male) at Howard University.	Guilford - Zimmerman Temperament Survey.	Members of a minority group do differentiate in the perception of the personality characteristics of their own group and of the majority group.

Brigham (1973)	To investigate the relationship between stereotypes and attitude, when stereotypes are defined in terms of criteria set by 3 very different samples of Ss.	114 black college students in predominantly black university. 86 white students in predominantly white university in same town in Northern Florida. 58 non-college whites from 2 small towns in Southwestern Georgia.	2 Attribution questionnaires ("own" and "others"). Short form of Multifactor Racial Attitude Inventory (1967).	Blacks' perceptions of the "typical" white student's racial attitude was closer to the expressed attitude of the rural noncollege whites than to the expressed attitude of the college whites. Number of stereotypes of blacks by whites was significantly related to racial prejudice for the noncollege whites, but not for the college whites. Evidence for a trait of stereotyping was found; Ss who expressed large numbers of stereotypes toward one ethnic group tended to do so toward the other ethnic groups also.
Brigham (1974)	To investigate the trait-attribution responses of both black and white Ss across a range of ages, in an attempt to provide evidence concerning patterns both between and within races in the attribution of "stereotypic-	501 white, 315 black children, from grades 4-12 in 2 schools in the Deep South. One school entirely white, one school entirely black.	Modified form of questionnaire developed by Blake and Dennis (1943) assessing views as to whether specified traits are more characteristic of "whites" or "Negroes."	"Regarded cross-sectionally, within-race agreement on direction of trait attribution increased with age in both samples, but whites showed more agreement than blacks at all grade levels" (p. 156). Largest amount of between-race agreement in attributions in grades 7-11.

Study	Purpose	Description of Subjects	Instruments, Methods	Major Findings, Author's Conclusions
	relevant" traits to blacks and whites for children of differing ages.		Approximately 1 year later, "trait-likeable-ness" questionnaire administered to 69 students in grades 6 and 7, and 79 in grade 11 in the white school.	White childred used "No Difference" response to a significantly greater extent than the blacks. Both samples showed a significant tendency to attribute favorable traits to their own race, but this tendency was significantly greater within the black sample.
Heiss and Owens (1972)	To provide an empirical test of hypotheses put forth by McCarthy and Yancey, concerning the self-evaluations of blacks and whites. (Basically, McCarthy and Yancey question the empirical, theoretical, and popular belief that blacks have lower self-esteem than whites.)	Data obtained from the two large-scale surveys conducted by the National Opinion Research Center in 1966. See Crain and Weisman (1972) in this table.	Instruments and Methods are as described in Crain and Weisman (1972).	The relationship between the self-evaluations of blacks and whites varies depending on the traits involved. For characteristics such as performance of family roles, no difference found between blacks and whites in the high-SES group, with a diffeence found favoring blacks in the lower-status group. For some other traits, whites did rate themselves higher in both SES categories. Authors stress the general similarity of the white and black subjects, and find no evidence that blacks are "crippled" by low self-evaluations.

Hraba and Grant (1970)

To duplicate Clark and Clark Doll Study in an interracial setting.

89 blacks, 71 whites, 4-8, randomly selected from public school system in Lincoln, Nebraska.

Procedures used by Clark and Clark followed as closely as possible. Ss interviewed individually using 4 dolls, (2 black, 2 white) by both black and white interviewers, and asked the following questions:

1. Give me the doll that you want to play with.

2. Give me the doll that is a nice doll.

3. Give me the doll that looks bad.

4. Give me the doll that is a nice color.

5. Give me the doll that looks like a white child.

6. Give me the doll that looks like a colored child.

7. Give me the doll that looks like a Negro child.

8. Give me the doll that looks like you.

The majority of children preferred dolls of their own race, regardless of color.

Black children in interracial settings are not necessarily white-oriented.

141

Study	Purpose	Description of Subjects	Instruments, Methods	Major Findings, Author's Conclusions
Mayovich (1973)	To compare the racial images of whites, blacks, and certain yellows (Japanese-Americans) as mutually perceived and differentially perceived by 3 different age groups.	100 white, 100 black, 100 yellow. 100 adults in their forties, 100 college-age, and 100 4th- and 5th-graders in each racial group.	Adjective list (Katz and Braly, 1933)—for the adults and college students. The children were asked to describe whites, blacks, and Japanese-Americans freely, in their own words.	"… The whites are portrayed as materialistic straightforward and pleasure loving; the Blacks are musical, impulsive and aggressive; and the Japanese, loyal to family, ambitious and intelligent" (p. 250). Hypothesis of mirror image (minority member's inauthentic acceptance of negative stereotype attributed to him by the dominant group) not completely supported. "The present study would seem to indicate that except among children, explicitly derogatory racial epithets are disappearing but are being replaced by more subtle ethnic humor" (p. 251).
Moore (1975)	To examine the usefulness of a feedback-based simulation paradigm in altering cross-racial attributions of personality characteristics.	30 white, 10 black male adolescents, grades 9-12, from a central high school in a Pennsylvania community (population 30,000 / white:black ratio = 90:10).	10 personality scales selected from Jackson's (1967) Personality Research Form. Black students served as criterion group. White students received Pretest, Training, and Posttest.	"In general, the results indicate that white male adolescents exposed to a two-session training progam can learn to simulate in a fairly accurate manner the self-description of black age peers on specific personality variables which initially showed dramatic differences in cross-racial trait attribution" (p. 237).

Author (Year)	Purpose	Sample	Method	Results
Polite, Cochrane and Silverman (1974)	(a) To examine the blacks' perception of blacks relative to whites, and (b) To examine a tenet of the black pride movement that more positive self-image will be associated with larger perceived differences between blacks and whites.	43 male, 53 female black undergraduate volunteers.	Semantic differential type questionnaire on which Ss were to evaluate the following concepts: "Me," "Blacks," and "White Protestants." Administered by a black investigator, completed anonymously.	Results indicated that: (a) the mean rating given "Blacks" was significantly more positive than the mean rating given "White Protestants" and (b) there exists a significant positive relationship between blacks' evaluation of "Me" and the extent to which blacks view "Blacks" differently from "White Protestants." Data suggest a change in self-image, and; the more positive a black individual's self-image, the greater is his perception of differences between the black and white Protestant groups.
Smith and Mazis (1976)	To explore (using unobtrusive measures) the relationship between self-designated racial labeling, and sex, age, income, job classification, and damage to dwelling.	Data obtained from files (1971-1973) of 145 tenants, who identified themselves as black, colored, or Negro, of a federally-funded, low-rent, public housing authority in a county in Northern Florida.	Data contained in each file: preliminary application form (later included verification of employment, income, welfare and/or social security); interview form; telephone contact sheet; lease; copies of all late rent notices.	There were significant differences between self-labeled blacks and non-blacks (colored and Negro self-labels combined) across the 5 variables: age, earned income, welfare, employment, damage to dwelling. Archival data supported proposition that unobtrusive measures of racial classification have psychological meaning.

Study	Purpose	Description of Subjects	Instruments, Methods	Major Findings, Author's Conclusions
				Not totally clear whether difference in self-labeling is associated with more positive self-identification among younger blacks or is indicative of flux of younger people adopting new language and/or identification with "black power" movement among youth. High correlation found between cluster of income-related variables and racial self-identification (indirectly related to self-concept).
Stabler, Johnson and Jordan (1972)	To measure attitudes towards self and others as associated with racial membership.	60 children (15 white males, 15 white females, 15 black males, 15 black females) selected from 4 integrated schools. Mean age = 5-9. Generally from middle- to upper-middle-income families.	Variation of Clark and Clark Dolls experiment. Statements emanating from speakers in different colored boxes, about which the Ss were to make evaluations.	"The basic prediction, that children would 'hear' the positive self-statements coming from the white box and the negative self-statements coming from the black box, was supported by the data … Attitudes toward the colors black and white may influence the way black and white children view each other and themselves" (p. 2097).

	Purpose	Sample	Method	Results
Teplin (1976)	To compare racial/ethnic group preferences of Black, Anglo, and Latino children, with focus on the latter—relatively little-studied—group.	398 (one-third in each group of Anglo, Black, and Latino) 3rd- and 4th-grade elementary school students in public school system of a large Midwestern city, from 12 classrooms in 3 different schools. All of the schools were racially/ethnically integrated; each of the 3 groups was in predominance at 1 of the schools.	Photo-choice method consisting of 24 pretested photographs, including 4 children of each racial/ethnic-gender "type". Ss were asked: "Who would you most like to have as a friend?" and "Who would you most like to work with?"	The majority of Anglos chose photos depicting Anglos; Blacks chose the photos depicting Blacks; Latinos did not choose photos of Latino children, but instead preferred the pictures of the Anglo children.
Ward and Braun (1972)	To investigate the interrelation of self-esteem and racial preferences.	60 black male and female children, aged 7-8. 30 from middle-class interracial suburban school, 30 from interracial inner-city school.	Piers-Harris Children's Self-Concept Test, read aloud to children by black female E. Adaptation of Clark and Clark Dolls test.	No significant sex differences. The majority of black children preferred the black puppet. Those Ss who made more black color preferences had higher self-concept scores than those who made fewer black color preferences. Definite change (from prior related studies) or relationship between self-esteem and racial preference "may signify a new spirit of dignity in the lives of Afro-American children."

(Group Membership—Racial and Socioeconomic)

Study	Purpose	Description of Subjects	Instruments, Methods	Major Findings, Author's Conclusions
Beglis and Sheikh (1974)	To look at black-white differences in self-concept, while attempting to control SES. Study undertaken for exploratory value, no prediction concerning outcome.	12 2nd-grade, 40 4th-grade, 24 6th-grade children attending two parochial elementary schools in Milwaukee, Wisconsin. Both schools of lowest SES level, with similar proportions of black and white students.	Modified form of Kuhn and McPartland's Twenty Statements Test. Given by regular classroom teachers.	Age is primary determiner of self-esteem. Verbally limited concrete concepts lead to more abstract self-perceptions. In all but two of the categories (Identification and Competence/ Task Achivement), there were no race differences. Main effects for sex not significant in any of the categories. "It appears that black children do tend to describe themselves more in terms of basic units of identification than do white children" (p. 109). (Authors recognize the importance of verbal ability and vocabulary needed for the particular measure used, and the possibility of an effect of added pressure by Ss' own teachers administering the test.)

Cicirelli (1977)

To examine the effects of SES and ethnicity in primary grade children, in order to provide recent evidence regarding this controversy.

166 male, 179 female, black and white, 1st-, 2nd-, and 3rd-graders, attending inner-city schools in a large metropolitan area.

Randomly drawn from 4 schools identified by school administration as serving a very low SES population; approximately 40% of the children were from families receiving welfare.

Purdue Self-Concept Scale, a revision of the Children's Self-concept Index used in Westinghouse study to evaluate Head Start children.

There were significant main effects of grade level and ethnicity—decline in self-concept with grade level, blacks scoring higher than whites.

Analysis of black 2nd-grade children's scores indicated that race differences were related to welfare status (resulting in higher scores).

There was a significant effect of SES level, with welfare children scoring higher on the self-concept scale than nonwelfare children.

The self-concept of nonwelfare blacks was found to be no higher than that of the white group.

Therefore, the original difference observed on the self-esteem levels between blacks and whites (blacks scoring higher) was found to be directly related to SES level (welfare status Ss scoring higher than nonwelfare).

Study	Purpose	Description of Subjects	Instruments, Methods	Major Findings, Author's Conclusions
Gestinger, Kunce, Miller and Weinberg (1972)	To evaluate the relationship between sociocultural variables and selected self-concept instruments.	198 metropolitan 6th-grade students, aged 11-14, in St. Louis. 55 black, from "Title I" schools, 47% male. 66 white, from "middle-class" school, 64% male. 77 white (1 black), from schools in wealthy section, 55% male.	The measures for self-concept: Coopersmith Self-Esteem Inventory, Soares and Soares Test, Ziller Test. The personal-social characteristics: SES, welfare status, race, educational advancement, sex.	". . . Findings indicate either that race and SES are inconsequentially related to self-concept in the age group studied, or that typical self-concept measures are not sensitive to differences that exist" (p. 149).
Healy and De-Blassie (1974)	To determine: (1) if differences exist in the self-concept among Black-, Anglo-, and Spanish-American adolescents, and (2) the extent to which these differences are influenced by ethnic group membership, socioeconomic position, or sex, or the interaction among these variables.	630 9th-graders (70% Anglo, 24% Spanish, 6% Black) in 2 junior high schools in the South Central New Mexico School System.	Tennessee Self-Concept Scale. Hollingshead's Two Factor Index of Social Position.	"Of the fourteen measures of self-concept assessed in this study, four scores were affected by the ethnic variable: Self Criticism, Defensive-Positive, Self Satisfaction, and Moral-Ethical Self" (p. 20). "Socioeconomic position was found to influence two of the fourteen scales utilized in this study, Social Self and Self Satisfaction" (p. 22).

| Roberts, Mosley and Chamberlain (1975) | To analyze age differences in "racial self-identity" in young children, the extent to which the child can identify himself as a member of a racial group. | 30 (aged 3-4, from 2 nursery schools), 30 (aged 6-7, from 2 elementary schools) black females, of middle-class parents, from predominantly black non-interracial urban areas of Washington, D.C. | Modified form of Clark and Clark's Dolls Experiment. | Significant relationship was found between age and accuracy of racial self-identity.

Racial self-identity is considered to be formulated during the preschool years for most black children and seems more firmly established by the beginning of the primary school years for most of these black children. |
|---|---|---|---|---|

The only scale of the 14 significantly affected by sex differences was Physical Self.

The Total Positive Score was not significantly affected by race, SES, or sex.

(See results in orignal source for more description of interaction effects on the scales mentioned above.)

Study	Purpose	Description of Subjects	Instruments, Methods	Major Findings, Author's Conclusions
Rosenberg (1972)	To learn how different social experiences, stemming from membership in groups characterized by differenct values, perspectives, or conditions of existence, would bear upon levels of self-esteem and upon self-value.	5,024 high school juniors and seniors from 10 randomly selected public schools in New York State, present on day of administration.	10-item Guttman scale containing self-descriptive statements, to which students responded "strongly agree," or "agree," "disagree," or "strongly disagree." Distributed to students by their teachers, in the classroom.	Only small difference between self-esteem of blacks and whites. "Self-esteem may be more a matter of one's position within one group than the rank of the group in relation to other groups" (p. 99).
Samuels (1973)	To investigate the self-concept of 4 race and class subgroups. Specifically, it was hypothesized that (as groups): (1) white lower class children would have a higher self-concept that black lower-class children, (2) that black middle-class children would have a higher self-concept than black lower-class children, (3) that white middle-class children would have a	93 kindergarten children randomly chosen from a heterogeneous population of 417 children attending a central school district in a New York City suburb. Sample was divided into lower- and middle-class according to education and occupation.	Clark U-Scale (1967). "Self as Subject" part of the Brown Test. Interviews with mothers at home before school year started (Aug., 1968). Children tested individually within first month of school, in a room set aside for testing purposes, for a total of 20 minutes for both tests.	The white lower-class children and the black lower-class children were not significantly different from one another in either test. The self-concepts of the black middle-class children were higher at a statistically significant level than those of the black lower-class children on both tests. This same class effect ws found for the white children, but only on the Clark U-Scale. On the Brown Test, there were no significant differences between classes, for the white children.

higher self-concept than white lower-class children, and (4) that white middle-class children would have a higher self-concept that black middle-class children.			No significant differences between the black and white middle-class children.
			Mother's church attendance and community group involvement were related to child's self-concept only for the lower-class children.
			No relationship between sex of child, nursery school attendance, intactness of the home, age of mother, mother's employment, or family mobility and the child's self-concept.
			Suggestion that class is more potent than race in the determination of self-concept.
Terry and Evans (1972)	To obtain independent measures of class and race discrimination, and to determine if these measures vary with race.	50 black, 50 white high school students from Louisville, Kentucky. Distributed to volunteers by study hall monitors.	Questionnaire composed of 20 descriptive statements for respondents to endorse.	Significant findings were that both black and white respondents attributed more discrimination, across both race and class lines, to other persons of either race than to themselves.
				Findings suggest that socioeconomic status (SES) may be a more important factor than race in affecting discriminatory behavior.

Study	Purpose	Description of Subjects	Instruments, Methods	Major Findings, Author's Conclusions
Trowbridge, Trowbridge and Trowbridge (1972)	To determine: (1) whether measurable differences in self-concept exist between children of different SES, (2) the dimensions of self-concept in which differences occurred, and (3) whether the differences in self-concept found in Trowbridge (1969, 1970) were confounded with other variables such as race, age, sex, and density of population.	3,789 students selected from 133 classrooms in 42 elementary schools, from both rural and urban central Iowa. 1,662 3rd- to 8th-graders with low SES, 2,127 with middle SES.	Coopersmith Self-Esteem Inventory. Administered by teachers in the classroom to entire class.	The higher self-concept of low SES students exists at all levels; any confounding of the observed SES results by I.Q. differences is minor. Higher self-concept scores were obtained for Ss who were low in SES, black, and/or from rural areas. Sex and age not significant factors. Further analysis revealed that low-SES Ss scored higher on General Self, Social Self-Peers, and School-Academic items, while middle-SES Ss scored higher on items concerning Home and Parents.

Author	Purpose	Sample	Instrument	Findings
Busk, Ford and Shulman (1973)	To investigate the extent to which the school experience affects a student's self-concept.	696 students in 6th-, 7th- and 8th-grades, from parochial schools in similar SES sections of Chicago. 160 from all-white school, 172 from all-black school, and 364 from 3 integrated elementary schools (53-59% black).	50-item Coopersmith Self-Esteem Inventory. 8-Item How I see Myself rating scale. Brookover's 8-item self-concept of ability measure. Tests given in class to everyone present on testing day (teacher not present). All 4 testers (3 female, 1 male) were white.	The more influential factor affecting students' self-concept is not ethnic membership, but racial composition of school attended. The self-concept of blacks does not differ from that of whites.
Carey and Allen (1977)	To evaluate the impact that participation in black studies has had on the self-concept and academic performance of black students on white campuses.	218 black students randomly selected from 5 historicaly white state universities in the Southwest.	Tennessee Self-Concept Scale. Extent of involvement in black studies and academic performance was determined by administration of a highly-structured, pre-coded, 49-item self-report Likert-scaled instrument developed by the investigator.	Participation in black studies programs was not found to have a significant effect leading to higher levels of self-esteem or higher grades than black non-participants. Author suggests that the early socialization experience is much more important to the solidity of the self-image—formed before high school and college.

Study	Purpose	Description of Subjects	Instruments, Methods	Major Findings, Author's Conclusions
Crain and Weisman (1972)	To determine the effects, if any, of attending integrated versus segregated schools. Developed into a larger study of Northern blacks, conducted by the National Opinion Research Center (NORC), at the request of the U.S. Civil Rights Commission, in the Spring of 1966.	1,651 men and women, aged 21-45, living in the metropolitan areas of the North—selected from 297 randomly chosen "city blocks" in 25 different areas. 6 black men and women in each block ("cluster sampling"). Another NORC survey (Amalgam Survey) included 1,326 white respondents, and was used as a basis for comparison.	2-hour interview containing nearly 500 questions and subquestions. A measure of self-esteem was obtained in the following manner: A list of 10 items was read and the interviewee was asked to rate himself average, above average, or below average on each. Of the 10 items, 3 focused on the person as a family member, 2 dealt with aspects of character, and the remaining items covered ability in 5 areas.	Book contains *extensive* description of results, covering many social, psychological, educational, and political issues. Focus here is on the results dealing specifically with self-esteem. Blacks found to have, on the average, lower self-esteem than whites, but these racial differences are not so straightforward. There is more "spread" in the distribution of self-esteem scores for blacks; blacks tend to have either very low or very high self-esteem. Blacks born in the South have generally lower self-esteem than whites, but blacks born in the North tend to have self-esteem as high or higher than whites. Authors conclude that this is not due to integration, but to the increased opportunity of expressing aggression in the North. Self-esteem associated with characteristic called "assertiveness," which is defined as functional

				aggressiveness (i.e., competitiveness that enables the individual to manipulate his environment successfully).
				Self-esteem also associated with dysfunctional aggressiveness (i.e., fighting and getting arrested).
				The critical factor is the individual's sense of internal control, which the authors find to be independent of self-esteem, at least for blacks.
				The data reveal 4 types of blacks who handle aggressiveness differently, depending on their self-esteem and sense of control.
			Data collected from case records of all black students who obtained counseling at the mental health clinic during the three academic years 1962-1972, to determine the proportion who presented complaints or symptoms of identity	Comparative content analysis revealed 4 distinct patters of coping behavior directed toward the resolution of the conflict generated by ethnic or socio-cultural marginality.
		22 females, 19 males, with modal age of 19.		Affirmation Mode—movement with the dominant culture.
Gibbs (1974)	To determine the types of responses black students employed in coping with identity conflicts and to ascertain whether or not certain patterns of response were related to specific social and psychological characteristics.			

Study	Purpose	Description of Subjects	Instruments, Methods	Major Findings, Author's Conclusions
	Describes patterns of adaptation developed by black students at Stanford in response to their minority-marginal status in a milieu which is predominantly white, upper-middle-class and suburban—based on case studies of black students who sought counseling at the college mental health clinic from 1969-1972.		conflicts or stress attributable to cultural adaptation. Cases were analyzed according to: SES of student's family; student's exposure to integration in high school; student's ability to handle academic tasks; and student's feelings of self-adequacy.	Assimilation Mode—movement toward the dominant culture. Separation Mode—movement against the dominant culture. Withdrawal Mode—movement away from the dominant culture. (Incidence of particular modes and their correlation with other variables is too lengthy to be discussed here but can be found in original article.)
Katz, Cole and Baron (1976)	To explore further the self-evaluative behavior and personality characteristics of male pupils in relation to their degree of academic success, using relatively large samples of boys, and black and white experimenters.	68 black and 88 white male pupils in grades 4-6 at 2 racially integrated small-city public schools in Michigan.	The Test Anixiety Scale for Children Reinforcement History Questionnaire. Intellectual Achivement Responsibility Questionnaire. Administered to groups of 15 or more.	A significant interaction effect was found between level of achievement and race of experimenter. When tested by a black E, high-achievers were more self-critical than low-achievers; when E was white, opposite relationship was observed—high achievers being less critical than low-achievers. Black pupils' perceptions of their social reinforcements from significant adults was found to be

Study	Purpose	Subjects	Procedure / Instruments	Results
			A few weeks after administration of above, Ss' self-evaluative behavior was tested individually.	significantly related to self-criticism. For the white Ss, the only relationships obtained were between measures of positive social reinforcement and self-approval. "In general, this study found more similarities than differences between black and white pupils on various motivational measures. There were no race differences in mean scores on achievement responsibility, test anxiety, perception of social reinforcement from significant adults or self-evaluation of social performance" (p. 373).
Martinek, Cheffers and Zaichowsky (1978)	To assess the effect of organized physical activity on the develpment of specific motor skills and self-concept of elementary age children from bi-ethnic backgrounds (black and white).	344 elementary age children, in grades 1-5, from the Boston public school system. Both groups (treatment and control) coming from similar social, cultural, and economic backgrounds.	KorperKoordination Test for Kinder (Schilling, 1974) used to measure motor development. Martinek - Zaichowsky Self-Concept Scale for Children. Piers-Harris Scale for Children.	Results strengthen the contention than an organized physical activity program has a positive effect upon the development of motor coordination skills in young children. No significant difference ws found in motor skill improvement between races (only between treatment and control groups).

Study	Purpose	Description of Subjects	Instruments, Methods	Major Findings, Author's Conclusions
			Coopersmith Self-Esteem Inventory.	An organized physical activity program may have a definite effect on a child's self-concept.
			The treatment group had a formal physical activity program for 45 minutes once per week, for 10 weeks.	The significant decline in the self-concept scores for both groups in Grades 3, 4, and 5 suggests that early school pressures interact with the development of the child's self-concept.
Payne and Dunn (1972)	To study the effect of group guidance upon the self-concept of culturally different students participating in an Elementary and Secondary Education Act, Title I, program.	45 4th- and 5th-graders in a consolidated school district in the Gulf Coast of Texas.	The Brown IDS Self Concept Referent Test (1966). The experimental group was treated to 18 specified group guidance activities for a period of 50 minutes weekly.	Indication that group guidance experiences are an important agent of change in self-concept, particularly in the component labeled "Self as Object," the way a person perceives how others see him. Treatment altered favorably the self-concept of the participants on referents "Mother," "Peer," "Self as Object," and total referent score. Little change in mean scores on referent "Teacher." "Overall, a comparison of the mean differences among races for the control group revealed that the

white-American began and ended with the higher 'Self as Subject' referent score, the Negro and white improved their 'Self as Object' scores, but the Mexican-American regressed in his perception of 'Self as Object'. Results for students experiencing group guidance were striking in the differences found. The Mexican-Americans showed the greatest amount of change in mean differences on the subtest 'Self as Subject', while the Negro-Americans and white-Americans made the greatest amount of improvement on subtest 'Self as Object' scores" (p. 162).

Powell and Fuller (1972)

Part of an ongoing study investigating the psychological impact of school desegregation on self-concept.

Specifically, that the amount and degree of self-concept change would be dependent not only on the degree of

1,754 white and black 7th-, 8th-, and 9th-graders in public and parochial, segregated and desegregated schools in 3 cities in the South, where social and political changes had either occurred in a relatively peaceful manner or with a great deal of conflict.

Tennessee Self-Concept Scale.

Ss interviewed by experimenter of same race.

Interviews with teachers, school administrators, and members of community.

(Only partial results available.)

Blacks scored higher in self-esteem.

Blacks in segregated schools had higher self-esteem than those in desegregated schools.

"It would seem from the data collected from the pilot study that being a black male in a predominantly black or all black school in a

Study	Purpose	Description of Subjects	Instruments, Methods	Major Findings, Author's Conclusions
	family stability and economic status, but also on the amount, degree, and kind of social, cultural and political changes within the community.			cohesive militant black community seem to be the major variables involved in positive self-concept. Secondarily, family stability and educational level of parents may also be important" (p. 526).
Shaw (1974)	To determine whether differences exist between disadvantaged black pupils and advantaged white pupils, and whether these differences change during the academic year.	146 black (67 male, 79 female), 545 white (254 male, 271 female) students from grades 2-6 in 6 elementary schools in a small Southern city.	18-item Harvey Self-Image Scale. Scale measured 4 aspects of a person's self-image; sociability, independence, hostility, and achievement orientation. (A second study with slightly modified design was also done to provide greater precision on some variables.)	The most significant findings (those that were replicated in both studies) were: (a) boys saw themselves as significantly less sociable but more independent than did girls (b) blacks perceived themselves as being significantly more independent and hostile than did whites (c) sociability generally increased and achievement orientation decreased as a function of grade level (d) pupils generally saw themselves as more sociable in the Spring than in the Fall, but

blacks decreased in sociability (Study 1) or showed no change (Study 2), whereas whites either showed no change (Study 1) or increased (Study 2) in sociability during the school year.

(e) pupils in grades 2 and 3 decreased in hostility during the school year, those in grades 4 and 5 showed no change, and pupils in grade 6 increased in hostility during the school year.

The sex differences observed were in accordance with societal roles.

Since the whites were generally advantaged and the blacks generally disadvantaged, the author sees findings as having some interesting implications for the controversy involving the self-perceptions of advantaged and disadvantaged youth.

Study	Purpose	Description of Subjects	Instruments, Methods	Major Findings, Author's Conclusions
Zirkel and Moses (1971)	To determine: (a) if differences exist in the self-concept among black, Puerto Rican, and white elementary school students, and (b) the extent to which these differences are influenced by the minority or majority status of each group within the school.	120 5th- and 6th-graders from 3 elementary schools in similar SES sections of a large Connecticut city. All three schools contained all of these ethnic groups, but with varying proportions.	Coopersmith Self-Esteem Inventory. Administered to small, ethnically mixed groups, away from classrooms, by multi-ethnic group of graduate interns.	Ethnic group membership in a public school significantly affects the self-concept of individual students; this seems to be attributable to presence, but not minority-majority proportions of ethnic groups within the school. Support given for the growing number of studies which indicate that the self-concept of black children does not differ significantly and may even be higher than that of white children. The self-concept of the Puerto Rican children was lower than that of both the white and black children.
			(Locus of Control)	
Garcia and Levenson (1975)	To examine the relationship between the multidimensional measures of locus of control and 2 demographic variables —ethnicity and SES.	194 college students (84 white, 110 black) attending one of 3 state universities in Texas. Most of the Ss came from a school in which their race was in the majority.	All Ss received an envelope containing locus of control scales, other personality tests and a number of demographic variables questions. The Internal, Powerful Others, and Chance	Students from low-income families had stronger perceptions of control by change than wealthier students. There were no significant class differences between groups on Internal and Powerful Others Scales.

Scales (Levenson, 1974) each consisted of 8 items presented to the Ss as a unified attitude scale of 24 items in a Likert-format (range on each scale: 0-48).

Blacks were shown to have higher expectation of control by Powerful Others than whites.

Blacks were shown to hold significantly higher expectations for chance control than whites.

Each S was assigned an SES level, based on parents' occupations using Hollingshead's Occupational Status Scale.

No significant difference was found on the Internal Scale between races.

Findings indicated that amount of income is related to perceptions of control by chance.

Authors suggest that these results be interpreted carefully because of the sampling procedures.

Kinder and Reeder (1975

To provide evidence on the organization of beliefs about personal control for a heterogeneous population.

To examine the internal consistency of the personal control dimension within representative

Data provided by 2 Los Angelos Metropolitan Area Surveys (LAMAS)—semi-annual omnibus survey carried out by Survey Research Center at UCLA—multi-stage area probability sample of approximately 1,000 persons.

Set of 4 personal control items from Rotter Scale was included in LAMAS II, conducted in Fall, 1970.

Parallel analysis conducted in Fall, 1971.

Unreliability of personal control factor dimension among blacks present within all age and education groups, regardless of interviewer involved, or sex of S; and it recurred in most respects in a comparable survey.

This unreliability was specific to the blacks.

163

Study	Purpose	Description of Subjects	Instruments, Methods	Major Findings, Author's Conclusions
	samples of 3 major ethnic groups in Los Angeles—Anglos, Blacks, and Chicanos.			Personal control dimension did show satisfactory internal consistency for the corresponding subsamples of Chicanos and Anglos. Theoretical and methodical implications of these findings are discussed, and ramifications for social indicator research noted.
Milgram (1971)	To obtain developmental data on Internal Locus of Control (ILC) across a broad age range of black and white children, and to correlate ILC with scholastic achievement, a non-scholastic intellectual measure, and an occupational aspiration measure.	80 children attending parochial schools in Washington, D.C. 20 each in Grades 1, 4, 7, and 10, each sample divided equally into male and female, black and white.	ILC scale developed by Bialer (1961). Metropolitan Readiness Test for Grade 1. Otis Lennon Mental Ability Test, for the other grades. Raven's Colored Progressive Matrices. Occupation aspirations based on asking children what they wanted to be when they grew up, rated using Roe Occupational Classification System (1956).	Age-related increments in ILC were found. There was an absence of difference in ILC between black and white children. (Authors note that these black children were Catholic, attending parochial schools and were not typical of the larger population of predominantly non-Catholic, black children attending public schools.)

Study	Purpose	Sample	Measures	Results
Page (1975)	To examine the relationship between self-esteem and belief in internal versus external control among 24 black youngsters participating in a summer aviation program at Tuskagee Institute, Alabama, in July-August, 1974.	21 boys and 3 girls (all black) from various geographic locations and income levels, chosen for program through various channels.	Personal data sheet (demographic information). Academic Achievement Scale (short version of WAIS). Combined internal-external control/self-esteem questionnaire (modified version of Rotter I-E scale, plus 10 items from Coopersmith Self-Esteem Inventory). All 3 completed at start of program. At end, 2 weeks later,22 SS completed combined internal-external/self-esteem questionnaire a 2nd time.	Highly significant correlations between academic achievement and self-esteem, and academic achievement and SES. Significant gain in self-esteem by Ss below 16 years old. Significant gain in self-esteem by Ss from middle-income families. Zero gain in self-esteem on part of Ss from low-income families. There was no significant increase in belief in internal control by any of the subgroups.
	To investigate the relationship among several sociological and psychological variables among a small population of black youngsters at the beginning of an exciting program, and to measure any change in the psychological variables at the termination of the program.			
Picou, Cosby, Lemke and Azuma (1974)	To investigate the types of occupational choices made by Southern black delinquent adolescents; to compare the occupa-	141 black 9th-grade males: 73 incarcerated at large all-black institution near large urban area in Louisiana; 68	Group-administered questionnaires (occupational desires and plans). Responses coded ac-	There were differences, though not significant ones, found between the delinquent and non-delinquent group. Goals, plans, and blocks were fairly similar between the two

165

Study	Purpose	Description of Subjects	Instruments, Methods	Major Findings, Author's Conclusions
	tional projections of delinquent and non-delinquent black 9th-grade males, and to determine the degree of perception of a set of factors as possible blocks to obtaining these occupational goals. Exploratory analysis, no formal hypotheses offered.	obtained from random sample of all-black junior high schools located in large urban area. Data collected April, 1969.	cording to modified version of the Alba M. Edwards (1934) socioeconomic classification of occupations.	groups, reflecting American cultural emphasis on occupational success. (Since race is not a variable, summary of findings will end here. See original source for more detail on differences between the delinquent and non-delinquent groups.)
Wax (1972)	To test the hypothesis that differences in self-concept might be found in a comparison of black and white pre-adolescent delinquent boys, in the area of future goal attainment. Specifically, that black children will have a lower concept of self and will feel less able to attain valued goals than their white counterparts.	20 black, 35 white institutionalized, male, pre-adolescent delinquents, aged 7-12.	Version of semantic differential technique (Osgood, Suci, and Tannenbaum, 1957) used to measure self-concept. 8 concepts: Men, Boys my age, Boys who get into trouble, I would like to be, I am, Boys who don't get into trouble, I will be, Women.	No significant difference between the black and white groups under study as to mean scores on the measure of self-concept. Only one concept attained a significant level of difference—Boys who get into trouble. The blacks' perception of this concept was as being a more positive characteristic than the whites.

(Father Absence, Role Models and Significant Others)

Study	Purpose	Method/Measures	Results
D'Andrade (1973)	To study the effects of father absence, specifically, the relationship between father absence and sex role idenfication.	121 children and 50 mothers from 58 black, working-class households in suburban areas outside of Boston. In 41% of the households, no father in residence.	The Franck Test (1949) —projective drawing test. Verbal self-description. Role-preference task. (The above 3 assumed to measure aspects of sex identity.) 15-minute observation of mother-child interaction in waiting room, through one-way mirrors.	A feminine response pattern was found on the Franck Test (unconscious sex role identity) for children who had not had a father present during their first 3 years, and also for children who had older sisters rather than older brothers. Presence of a father was also found to be an important influence on the measures of conscious sex role identity and sex role preference. This influence was conceived to be a function of the process of reciprocal role learning and perception of advantages and disadvantages based on sex role, rather than through the process of early identification.
Darden and Bayton (1977)	To study the relationship between high and low self-concept black men and women and their assessment of black male and female leading roles in motion pictures and television situation comedies.	40 male, 40 female black undergraduates at Howard University, aged 17-28. (This group represented the 20 highest and 20 lowest scoring males and females on the self-concept measure.)	Tennessee Self-Concept Scale. Stereotyping procedure (adjective checklist) to determine assessments of the leading black roles in movies and television.	Results showed significant interaction of all the factors, and mediating variables in the design. (They are too lengthy to delineate here; please refer to original source.) In general, it was found that black self-concept is a factor that mediates in assessment of black leading roles in the media.

Study	Purpose	Description of Subjects	Instruments, Methods	Major Findings, Author's Conclusions
				"The manner in which this variable mediates is a function of the sex of the blacks making the segment, the medium (motion pictures or television), and the sex of the leading roles being portrayed" (p. 623).
Earl and Lohman (1978)	To examine whether latency age black male children from homes with absent fathers have access to their fathers or to other males who might serve as role models.	53 latency-age black boys (between 7-12) in Knoxville, Tennessee, from low- and middle-income homes. Randomly selected from 3 subsamples, membership in a large, predominantly black church, population of 2 federally-supported housing projects. Mothers were also interviewed (N = 39). In all cases, the father was absent from the home: in most cases, the father had been absent for four or more years.	Separate interviews were conducted with mothers and sons, concerning boys' contact with their fathers and other adult males.	For half the boys, their father was not a distant figure, but someone they saw with reasonable frequency. All the boys in the study had access to some black male who could, at least potentially, serve as a role model. Appears to support claim by Chestang (1970) and others that males with whom black children can relate are present in both the black family and community.

Hartnagel (1970)

To explore the question of the effects of father-less homes; part of a larger investigation into conceptions of self held by adolescent males.

275 inner-city male high school students in urban area of Indiana.

School was of mixed racial composition.

Semantic differential qustionnaire containing a number of separate adjective dimensions.

Each dimension formed by pair of adjectives (for example good-bad) with a 7-point scale separating them. For each pair, S instructed to describe himself by checking position along scale.

Questionnaire administered to small groups (10-20) in separate room.

All categories of adolescent boys studied exhibit significant discrepancies between their actual and normative self potency.

(Based on idea of taking others attitudes about oneself and thereby constructing a self, "normative self" refers to attitudes of others with respect to what one should be, and "actual self" refers to attitudes of others with respect to how well one measures up to these normative expectations.)

These differences do *not* appear to be a simple effect of the variable of father availability.

Black father-absent boys had significantly less discrepancy between actual and normative self potency than white father-absent boys.

No significant difference for father-present boys.

Study	Purpose	Description of Subjects	Instruments, Methods	Major Findings, Author' Conclusions
Hunt and Hunt (1975)	To explore views regarding the structural circumstances of father-absence effects by comparing the orientation toward conventional achievement and personal identities of white and black adolescent males from father-present and father-absent families.	Data obtained from a large scale investigation of the self-images and perceptions of school-children conducted by Rosenberg and Simmons (1971, 1972). This sample was comprised of 1,917 students enrolled in grades 3-12 of the public school system of Baltimore, Maryland during the Spring of 1968. For this investigation 445 male students enrolled in the junior and senior high schools were selected from the original data base.	Data extracted from Rosenberg and Simmons' sample—the following variables were examined: Father Absence Social Class Orientation Toward Conventional Achievement. Personal Identity: Esteem and Sex Role.	**Results indicate that father absence has quite different consequences by race.** **Authors report some achievement and identity "costs" for white boys, and the possibility of some positive identity "gains" for black boys.** **This pattern was consistent across the social class levels.**
Taylor (1977)	To examine the personal aspirations, values, and significant others of male and female black college youth (particularly those now attend-	162 black students (46% male, 54% female) attending a large New England state university, aged 18-23, coming large-ly from working class	Ss were asked to rate, on a 5-point scale, 12 interests and activities according to degree of importance (and relative importance) he or she	**"In both normative content and choice of significant others, the modal value orientation of our sample of black college youth is observed to give high priority to traditional achievement values;**

ing predominantly white colleges and universities).

and low-income families.

expected them to have in life after graduation.

To identify the orientational others and role-specific significant others, 2 open-ended questions were employed, both a slightly modified form of those used by Denizen (1966).

("The term 'orientational other' was introduced by Kuhn (1964) to refer to those others with whom the individual has a history of relationships in contrast to social others in interaction with whom the individual's relationships tend to be more role-specific or situationally determined" (p. 798).)

those pertaining to hard work, success, material comfort, and self-actualizing" (p. 807).

Study	Purpose	Description of Subjects	Instruments, Methods	Major Findings, Author's Conclusions
Teahan and Po-dany (1974)	Resarch stemmed from the observation that the absence of good models within the immediate environment of lower SES black youth influences a whole range of values and self-concept.	109 black students from an all-male Catholic high school, in grades 9-11.	Modified version of Photographic Technique used by Secord (1959) in studying prejudice among whites. Social Survey Questions —primarily composed ot items taken from the Levinson F-scale.	"Films of successful blacks seem to have differential impact on higher and lower socio-economic black youth. While higher socio-economic students who saw the film reacted more positively than their controls to white photographs, there were trends in the opposite direction for lower socio-economic experimentals. There was also felt to be some evidence that an increase in pride about race may result in more negative or critical attitudes toward whites, especially among [lower] socio-economic students. It was suggested that these anti-white statements may be temporary and represent the first step in the development of a more positive self-concept among blacks" (p. 279).

Bolling (1974)

A preliminary report of a larger study examining self-concept of black children from different national and cultural backgrounds, and the relationship between self-image and depressive reaction.

This paper limited to presenting validity statistics on the measure, "The Black Identity Test".

711 black, 192 Puerto Rican, 111 white Puerto Rican, 189 white, male and female children, aged 3-17, from public and private schools in the New York metropolitan area.

Black Identity Test—scale developed by the author, which provides a measure of black identity for black Ss, aged 5-17.

(Test being validated against number of criteria.)

Blacks obtained highest score, followed by black Puerto Ricans, white Puerto Ricans, and whites.

Results point to 3 important facts:

(1) test has no value below the 5-year level

(2) black identification increases with age, and

(3) clear sex differences exist in degree of identification made by males and females.

Brand, Ruiz and Padilla (1974)

To understand the dynamics involved in respondents' decisions of ethnic identification and preference — reviewing and analyzing the varied research methods and the contributing variables influencing the respondent's choice.

Reviewed empirical investigations of ethnic identification conducted in the U.S. since 1900.

Categorized 8 basic designs for investigating ethnic attitudes and preferences:

(a) survey of verbalized attitudes as measured by ranking scales.

(b) preferences of photographs or line drawings of individuals from varied ethnicities.

Instrumentation and design effects were found, suggesting that contrasting (even similar) results be interpreted in terms of the above categories.

"With the range of methodology, control, and populations explored in childhood ethnic research, a holistic compilation of results is impossible" (p. 883).

Study	Purpose	Description of Subjects	Instruments, Methods	Major Findings, Author's Conclusions
			(c) choice of dolls of varied skin and hair colors	The assumption that ethnic preference can be evaluated as a single construct seems to be oversimplified.
			(d) cross-ethnic comparisons on personality assessment devices	Multiple measures of ethnic preference should be obtained from respondents.
			(e) analysis of sociometric interaction	A host of ethnicities besides whites and blacks should be investigated to determine ethnic specific and universal factors associated with ethnic esteem.
			(f) observation of intergroup behavior	
			(g) attitude basis in disguised measures	Description and/or control should be established for those factors that appear to influence strongly the respondent's ethnic identifications and preference.
			(h) measurement of autonomic changes	
			9 major variable isolated and examined for their role in studies of ethnic identification and preferences:	
			(a) examiner ethnicity and sex	
			(b) geographical residence of target population	

(c) population proportion of each ethnic group in the research sample

(d) length and quality of contact

(e) subjects' socioeconomic level

(f) subjects' ages

(g) sex of ethnic stimuli employed in instrumentation

(h) subjects' sex

(i) intensity of skin color of subjects and of ethnic stimuli employed in instrumentation

Study	Purpose	Description of Subjects	Intruments, Methods	Major Findings, Author's Conclusions
Burbach and Bridgemen (1976)	To examine the relationship between Coopersmith's Self-Esteem Inventory and the Intellectual Achievement Responsibility Questionnaire, centering on black and white 5th-graders.	274 5th-grade students from 2 urban Virginia public grade schools. (No direct SES measure obtained—SES and race may be partially confounded.)	Coopersmith Self-Esteem Inventory. Intellectual Achievement Responsibility Questionnaire. Group administered during same testing session.	In general, a lack of relationship was found between the two variables. Thus, high self-concept is only slightly related to accepting responsibility for success, and except for white males is unrelated to a willingness to accept blame for failures.
Jones (1978)	To explore black-white personality differences and to meet objections that were raised in connection with previous work by: (a) investigating a sample population of normal young adults, (b) employing personality instruments that are not oriented toward pathology, and (c) abandoning the use of misleading scores on scales that have been standardized on white populations in favor of an item level analysis as a	226 black and white, male and female, junior college students in the San Francisco Bay Area.	An early version of the Haan (1965) coping and defense scales designed to measure ego processes and derived from both the MMPI and the California Psychological Inventory. Embedded Figures Test (a perceptual-cognitive performance measure).	The most striking finding was the sheer magnitude of differences between races, variance of this degree not previously reported. Even when SES and education level are controlled for, important race differences remain evident. (Please refer to source for detail.) The young black appears not as an estranged victim, but as a balanced, differentiated personality—not alienated from society; but, psychologically speaking, very much in the system.

more direct avenue for a better understanding of the *meaning* of black-white personality differences.

Sappington and Grizzard (1975)

To examine the effects of the presence of a white experimenter on the performance of blacks.

To determine if performance decrement (decrease in the performance of blacks on intellectual tasks in the presence of whites) still exists.

40 black male students in 7th- and 8th-grade at a junior high school in Lexington, Kentucky, where student body was overwhelmingly black and faculty was racially mixed.

The Digit Symbol sub-scale of the WAIS.

Subject attitude scale.

Test battery of various personality variables including:

Manifest Anxiety Scale (Taylor, 1953).

The Test Anxiety Scale (Sarason and Ganzer, 1962).

Manifest Hostility Scale (Siegel, 1956).

Minnesota Multiphasic Personality Inventory K Scale (Dahlstrom and Welsh, 1960).

Black Ss were found to respond differently depending on whether they were supervised by whites or blacks—performance *increment* was found (in presence of white Es), even when task was described as intellectual.

177

Study	Purpose	Description of Subjects	Instruments, Methods	Major Findings, Author's Conclusions
Schofield (1978)	To examine the possibility that the Draw-A-Person Test may serve as a useful indicator of the acceptance of racial identity, and to explore its reliability and validity as such a measure.	157 black and 167 white 1st- and 2nd-grade children (approximately equal numbers of male and female) attending 4 public schools in lower-class and working-class neighborhoods in a large Northeastern city.	Draw-A-Person Test, class administered by teacher and investigator. 28 black children retested 5 weeks later, 6 identification tasks added to drawing.	Race of figures drawn by black children was found to differ significantly from those drawn by whites, blacks tending to draw blacker figures. There were no other significant main effects or interactions. Concluded that the Draw-A-Person Test does tap acceptance of racial identity; this was suggested by the significant correlation between the race of the figure drawn and a more traditional measure of racial self-identification.
Zirkel and Gable (1978)	To examine the test-retest reliability and construct validity of selected types of self-concept measures: nonverbal, verbal, pictorial, and observer ratings, for black, Puerto Rican, and white adolescents.	218 (45 black, 132 Puerto Rican, 41 white/approximately equal number male and female) 7th- and 8th-graders from 10 classrooms in 2 inner-city public schools in large Connecticut city.	Coopersmith Self-Esteem Inventory. Primary Self-Concept Scale. Self-Esteem subtest of the Self Social Symbols Tasks (SSST). The Behavior Rating Form.	Except for the SSST, most indicators were found to be somewhat stable over the 3-week period (some ethnic differences were found in the stability reliabilities). "The reliability and validity data reported in this study indicate that evaluations of self-concept enhancement programs for different ethnic groups should give considerable attention to selection of evaluation measures" (p. 51).

Teacher Rating Form.

Tests administered, on test-retest basis over three-week interval, to groups of 15-25 pupils, by 3 Education graduate students, each representing one of the three ethnic groups in the study.

CHAPTER 9

Theoretical Issues

There are a number of important issues within more conceptual writings which are relevant to our discussion of identity formation. In this chapter we first address one more aspect of the empirical research. In subsequent sections, we take up several related themes in the literature. These themes include identity formation as well as the broader area of research in black personality development.

DEVELOPMENTAL STUDIES AND LONGITUDINAL DESIGN

One of the most striking observations of our previous chapter is the paucity of longitudinal studies directly addressing identity development. Most of the studies conducted over the past ten years have looked at personality variables such as self-concept or self-esteem, and have used cross-sectional designs. In contrast, our original study focused upon identity formation (in terms of self-images) and relied upon a longitudinal design. There have been a number of quasi-longitudinal studies, where children at different ages are compared; but no studies comparing *change* over time are reported. This represents an important omission and perhaps reflects the general difficulty of carrying out longitudinal studies, compounded by the additional complex problems inherent in performing racial studies.[1]

We see identity formation as a developmental process, an intertwining of intrapsychic and social forces, changing over time. As was stressed earlier, to study development requires "time-based comparisons of...levels...Observations from a single point in time cannot by themselves adequately characterize this developmental process. It is a patterning over time of particular variables that serves to define the type of identity taking place" (p. 30). Based on Erikson's definition of "...an evolving configuration...," the single-shot measurement

[1]Many of the problems are discussed by Hauser in *Racism and psychiatry: Thinking about race and racism: Clinical and research dilemmas*, (1973).

method is simply not adequate. Other writers (for example, Damon, 1979 and Wohlwill, 1973) have also discussed the importance of longitudinal versus cross-sectional methodology when studying developmental change.

This is not merely a question of research style; it is also a conceptual issue, in that one's psychological perspective is intrinsically related to all aspects of the research endeavor. What we want to point out is that there is a major difference between the types of studies reported over the past ten years and our initial study of identity formation. We see these many studies as focusing on "aspects" of a much larger picture (e.g., self-esteem as an aspect of identity formation), representing one facet of a more complex configuration.

We do not expect researchers in this area to adopt a developmental perspective, together with a longitudinal approach. Our position, however, is that identity formation is best assessed through both empirical, and clinical studies based on many observations over time, rather than through single observations of individuals which *may* indicate single points of the developmental process. This is especially true in studies of adolescents, this period being "... an especially difficult time for pinning down any lasting and positive self-identity" (Maldonado, 1975). Single testing would be likely to produce a distorted picture.

INTERTWINING OF IDENTITY ISSUES

Even when writers employ the same term, this does not necessarily mean that they are referring to the same thing. One distinction that we found in our review was that of group (or racial) identity versus personal identity.

Personal identity and group identity are viewed by some as two aspects of identity, whereas others see one as contained within the other (e.g., group identifications as part of identity formation). In other words, there are varying degrees of emphasis concerning the impact of social dimensions upon the formation of identity. Three positions can be identified in this area:

(1.) There are those who are basically concerned with *personal identity*, though they do differentiate various aspects, viewing group membership (or identification) as a part of identity development. This approach (exemplified in our original study and by Erikson) focuses "... on the more individual aspects of the problem, as in psychoanalysis" (p. 5). Sociocultural influences are considered to be important, but not the primary factors to be considered in investigating identity formation. The "personal identity" (or ego identity) view is discussed more extensively in our first chapter.

(2.) In contrast there is a more social-psychological approach which places strong emphasis on the group aspects of identity. Such a perspective ". . . regards the individual adolescent as a member of society, and considers his actions, his problems, even his aspirations, to be the resultant of forces which have their origin in social developments" (p. 9).

Examples of this approach, in which identification with a group is considered a major element in the establishment of identity include: Goldschmid, 1970; Dizard, 1970; Porter, 1971; Goodman, 1972; Gurin and Epps, 1975; Taylor, 1976b; and Rosenberg, 1972, 1979, among others. The importance of individual personality is recognized by some of the discussions. Comer and Pouissant (1975) note that a positive group identity is also dependent ". . . on an inner core of pride and positive feelings (primary identity) . . ." On the other hand, Proshansky and Newton (1973) make one of the strongest arguments for the impact of group identification upon identity, seeing ". . . racial group membership the nexus of . . . emerging self-identity . . . a negative self-identity is frequently rooted in negative group identification. From this assumption, we would expect the converse to follow: that positive self-identity is dependent on positive group identification. Considerable evidence supports the idea that personal or self-pride is essentially the expression of group pride" (p. 181; p. 207).

(3.) Finally, there are those who focus exclusively upon group membership, and who are specifically interested in the identification aspect of identiy. A number of empirical studies have been done attempting to tap the nature of *racial* identity. For example, Schofield (1978) used the Draw-a-Person Test as an indicator of acceptance of racial identity. Roberts and colleagues (1975) looked at age differences in "racial self-identity"—the extent to which children were able to identify themselves as members of a racial group. And Bolling (1974) designed a scale, the Black Identity Test (BIT), to provide a measure of black identity for black subjects, ages five to seventeen. (For more details about these studies and their findings, see the Table in Chapter Eight.)

One must carefully attend to these variations when attempting to compare results between studies. As Krate (1974) points out: "In one respect, Hauser's findings may not be directly compared with ours in that, relative to our Q-sort items, his may have tapped personal identity more than racial or ethnic identity, whereas, by comparison, the present study may be tapping more of the racial rather than the personal components of development of identity" (p. 1075).

It is also important to be sure that the measure being used is assessing the aspect of identity one is indeed interested in. Maldonado (1975) comments upon the confounding of personal and group identity, "To claim that a test measuring individual characteristics is a reflection of group characteristics is a misuse of research" (p. 619). This confounding is of course reflected in the opposite fallacy as well, "The problem concerning the level of analysis and the level of application is also involved when general group characteristics are used to explain and predict individual or personal characteristics".[2]

SIGNIFICANT OTHERS AND ROLE MODELS

Significant Others

Significant others and role models have a powerful impact on identity formation. Significant others are seen, especially in the social psychological paradigm, as those who shape our personalities through their evaluations of us and the feelings we have resulting from those evaluations. Many of the ideas in this area are based on the works of Mead (1934) and Cooley (1902). Mead, in his symbolic interactionist model, basically characterized the self as being a process lodged within interpersonal interaction. Cooley viewed the social self as a looking-glass, or mirrored, image of who we are. In other words, what others think of us is reflected back, thus forming our concept of identity.

These ideas are still popular today. Pettigrew (1970) expresses Cooley's notions exactly: "We learn who we are and what we are like by observing how other people react to us." For a number of years, whites have been considered to be the significant others for blacks, reflecting back to blacks, through racism, a negative derogatory image. This has been used to explain the negative self-concept or lowered self-esteem (supposedly) held by many blacks. Blacks were considered to acquire their identity through interaction with significant others—whites being "the significant others in American society" (Banks, 1972).

More recently, however, this conceptualization has been seriously challenged. There are many who accept the argument that the self does emerge through interaction with significant others, but doubt that whites are significant others for blacks (Baughman, 1971; McCarthy and Yancey, 1971; Taylor, 1976a; Taylor and Walsh, 1979; Rosenberg, 1979.) It is believed rather that the significant others of most black

2Maldonado, 1975, p. 619.

Americans are fellow blacks rather than prejudiced whites; that blacks choose reference groups within the black community to provide social comparison. Taylor and Walsh (1979) express this position in concluding that, "the life styles and opinions of persons in one's immediate environment are much more important for most people than some presumed standards of the larger society" (p. 251).

At issue here is the question of who are the significant others. Taylor (1977) offers a distinction in the area of significant others: "The term 'orientational other' was introduced by Kuhn (1964) to refer to those with whom the individual has a history of relationships in contrast with social others in interaction with whom the individual's relationship tends to be more role-specific or situationally-determined" (p. 798). This distinction is made to call attention to the quality and intensity of the relationships with significant others and to the specificity and diffuseness of their influence. Some people may be significant in different ways and for different reasons—with different effects.

Another unsettled problem is whether the derogatory self-definitions are internalized to the extent that many have suggested. Baughman (1971) argues that blacks have resisted these negative definitions to a large extent. Their self-esteem has its first roots in the black community. Black children, in this view, confront white values and definitions *after* the foundation of their self-esteem has already been established within the black community. This is an important point to keep in mind, in relation to the ages of subjects in the various studies—to what extent they have come into contact with the larger (predominantly white) social institutions. A related assumption, also open to question is that there is direct relationship between disadvantaged social position and diminished self-regard. Such a viewpoint fails to take into account the multiple sources from which self-regard may be derived.

Role Models

The discussion of significant others emphasizes the social aspects of identity formation in imposing attitudes upon a person. Consideration of role models looks more closely at the individual, in terms of choosing whom he or she will emulate. In other words, we can distinguish the focus on significant others as representing a situation-centered perspective (or interactionist), while the focus on role models is more characteristic of a person-centered perspective. As described in Chapter 8, previous writers have commented on the limited range of available role models for blacks and how this might affect the development of identity.

Who are the (positive) role models for blacks? The debate on this issue is similar to the one concerning significant others. Some have felt that only blacks can serve as role models for other blacks. "Henderson (1967), for one, has taken issue with the notion that only black people can serve as effective work role models for minority youth. He emphasizes that such an approach is unrealistic because it denies the potential positive influence of a vast reservoir of white role models in occupations wherein blacks are either in small numbers or nonexistent" (Smith, 1975, p. 43).

Goldschmid (1970) agrees that whites can be models for blacks, but for a different underlying reason. "In addition to black leaders, whites have served as role models for those large segments of the black community that perceive in their adoption of middle-class values and behaviors an escape form the stranglehold of the ghetto life and a promise that they can achieve security and respect" (p. 19).

There is also a problem of how we are to define "positive"—black or white. There are individuals in black communities who may not fit any of the mainstream cultural standards of success; yet they function most competently in both black and white worlds. One need only look back through black history to find individuals who evidenced strength, courage, and pride, even when they were not free, surviving in severely adverse conditions. In the literature, and our original study, too little attention was given to these less obvious heroes. What appears at first as a picture of a few positive heroes, may, on more intense exploration, turn out to contain important cultural heroes, whose meaning was initially not so clear.

THE BLACK FAMILY

The black family is portrayed as being an important ingredient in the formation of a positive identity. Carter (1972) comments upon the importance of understanding one's history; black Americans must be taught to appreciate their unique heritage. He sees the acquisition of a more acceptable identity for blacks as beginning in the home, with the parents, and in the public schools. Comer and Poussaint (1975) also stress the developmental importance of learning black history and experiencing black culture in their comments on, "helping a child to learn strategies to deal with racism and the negative feelings that racism incurs."

Other recent discussions take issue with stereotypical portrayals of black families. One prominent stereotype, maintained over many years, is the description of the black family as "matriarchal," or at the very least "mother-centered." The Moynihan Report (1965) further perpetuated the simplified picture of black families,

In essence, the Negro community has been forced into a matriarchal structure which, because it is so out of line with the rest of American society ... imposes a crushing blow on the Negro male and, in consequence on a great many women as well.[3]

A corollary to the new seriously challenged "matriarch" thesis is the picture of black families as having "absent male authority figures." Just as the matriarchy conception is no longer tenable in light of empirical and historical evidence, so too the closely connected male role model "deficit" characterization must be dismissed as an accurate description of black families.[4] A number of studies question the assumption of the "damaging" developmental effects wrought by father absence in black families. Earl and Lohmann (1978) studied families where the father was absent. They found that for half of the boys, their father was not a distant figure, but someone they saw with reasonable frequency. Furthermore, all the boys in the study were found to have access to some black male who could, at least potentially, serve as a role model. Earl and Lohmann understand these observations as supporting the claim made by Chestang (1970) and others that males with whom black children can relate are present in both the black family and community.

The findings of Kellam and colleagues (1977) vividly convey the need to carefully describe black families prior to general conclusions about their structure or impact upon development. Looking at family arrangements in a poor black urban community and defining the families in terms of the adults present at home, they found eighty-six different family types. Such a result is consistent with the view that two-parent nuclear families are an "ideal" inappropriately used as a referent in research studies. No homogeneous entity as "the Black Family" exists. Billingsley's (1968) *Black Families in White America* develops similar points with respect to the heterogeneity of black families.[5] The major point emphasized by these studies is the importance of rejecting the simplistic stereotypes about "the Black Family" which detail its "tangle of pathology," "disorganization," "father-absence," or "matriarchal structure."

This brings us to a discussion of a different type of model—

---

[3]Quoted in Thomas and Sillen (1972), p. 87.
[4]The mounting arguments against both these stereotypes, and others about the black family are well summarized and discussed by Thomas and Sillen (1972).
[5]In addition, Billingsley also discusses the "successful" black families, their histories and sources of strengths.

conceptual models. In recent years, several writings have presented alternative approaches and models for understanding black families and black personality development. These paradigms have important implications. For example, they highlight subtle trends of racism, and other biases (medical disease view) embedded in previous modes of interpretation. Also, viewing various models used to understand the same phenomena provides us with a lesson in perspectives: how tied our interpretations of results are to our prior conceptions, and to the explanatory frameworks prominent in the field at a given time.

THE DEFICIT/PATHOLOGICAL MODEL

Previously discussed by Frazier (1939), Kardiner and Ovessey (1951), Moynihan (1965), Rainwater (1966), Grier and Cobb (1968), and more recently by Banks and Grambs (1972) and Proshansky and Newton (1973), this model views blacks as the vulnerable, psychologically crippled victims of white racism—due to the devastating, lasting effects of slavery, as well as to general existing negative conditions. The forces of racism, oppression, and discrimination are seen as having victimized blacks into hating self (blackness) and desiring whiteness (Kardiner and Ovessey). Grier and Cobb describe the consequence of unexpressed anger and aggression toward whites as turning inward and developing into self-hatred. This model is frequently referred to as the "American Dilemma."

Often this orientation has had social policy implications (for example in the writings of Frazier and Moynihan), pointing to the "disorganization" of the black family and the general "pathology" of black personality development. Blacks are compared to some "ideal" presumably found in whites. Not surprisingly, on this basis the blacks show deviant developmental patterns. One of the most powerful implications of this model is the near impossibility of healthy development for blacks. Historically, as well as logically, this seems to be one of the most serious flaws. The interpretation of the original results of our initial study was clearly influenced by this model which was a far more prevalent mode of explanation at the time.

THE FUNCTIONAL/ADAPTATIONAL MODEL

The Functional/Adaptational Model is one response to the serious problems presented by the Deficit/Pathological perspective. There are two basic emphases in this paradigm. First, patterns described in the Deficit Model are considered as adaptations to realistic conditions, and hence, are actually functional on the part of blacks. (For example, identity foreclosure could be viewed as a realistic way of coping with the limited opportunities available.[6] The second emphasis looks to African culture for roots and heritage, and links many black (Afro-American) practices and beliefs to African influence. This perspective, appearing in the literature as early as the late 1960's, can be found in Billingsley (1968), Valentine (1968), Davidson (1969), Blassingame (1972), Hill (1972), Nobles (1972), and Staples (1972). It has attained increasing prominence up to the present time.

A major theme expressed in these writings centers upon the idea of survival, of adaptation to "negative" surroundings. "Thus, the American Negro's survival in America can be summed up by stating that he has the ability to adapt; and that this ability had come down through the generations of blacks in America right from the first slaves brought from West Africa, where adaptability is one of the prime requirements of life" (King, 1976, p. 166). Other expressions of this model can be found in Comer and Pouissant (1975). "The black child has been forced to live in two cultures—his own minority culture and the majority one. He has had to teach himself to contain his aggression around whites while freely expressing it among blacks. Some people call this a survival technique" (p. 19). Peters (1978) also calls attention to coping patterns, "Behaviors previously viewed as *maladaptive* or *pathological* because they *differed* from the mainstream culture are interpreted *functionally*; for example, as having *survival value* within the Afro-American culture" (p. 656). In the same vein, Scott (1970) argues that the behavior of lower-class black adolescents should be viewed as meaningful and utilitarian behavior within the context of the life style of these individuals, rather than appearing from the perspective of middle-class values to be illogical and irrational.

Another important point noted by writers from the adaptation perspective is that in previous analyses (based on the deficit model) blacks were usually compared to an "ideal"—an analytic construct that does not fit most white families either (Staples, 1974; McCarthy and Yancey, 1971). "One thing is certain: the nuclear family of reality is significantly different from the *ideal* nuclear family projected by social science fiction" (Sudarkasa, 1975, 0. 293). Instead of citing its deficiencies from the ideal case, the black family should be viewed as a readaptation of African family structure, with roots tracing back to traditional Africa (King, 1976; Nobles, 1978).

Jones (1979) sums up these issues in calling attention to the failure of psychology professionals. Black (African-American) clients are not

---

[6]Jones' (1979) description of the social atmosphere surrounding black children suggests that what we previously termed "identity foreclosure" may be an accurate appraisal and functional coping behavior on the part of black adolescents.

being examined in terms of their own culture, resulting in distorted diagnoses, based on white culture and society. It is necessary, she argues, to examine African-American culture in terms of roots and practices which are traceable to traditional African culture and to the *functional* quality of blacks' experience in America where "...survival needs [are] an ongoing preoccupation..."[7]

Valentine (1971) presents arguments against the prevailing interpretations of Afro-American culture, and constructs what could be viewed as a third conceptual model. He "proposes a bicultural educational model, recognizing that many blacks are simultaneously committed to both black culture and mainstream culture, and that the two are not mutually exclusive as generally assumed." Valentine sees the concept of biculturality as "the essence of the divided entity symbolized by the very name Afro-American." ... "The central theoretical weakness of the 'difference model' is an implicit assumption that different cultures are necessarily competitive alternatives, that distinct cultural systems can enter human experience only as mutually exclusive alternatives, never as intertwined or simultaneously available repertoires" (p. 5).[8]

BLACK PSYCHOLOGY AND THE RELEVANCE OF RACE

These various models bring us to the discussion of another important issue, one which is also vigorously debated in the literature— whether or not there is or should be a separate Black Psychology.

Comer and Pouissant assert that "there is no such thing as a black psychology or white psychology." They emphasize that psychological practices in the United States have been white-dominated, often culture-biased and racist, and that psychological tests have been misused; these are abuses which must be corrected. "Nonetheless, blacks and whites alike experience aggression, affection, independence, and dependency. Everything the black mind is capable of, the

[7]Related to the adaptation model is the "difference approach" advocated by Yando, *et al.* (1979), "in which no group is considered inferior or superior to any other and in which differences among groups are viewed instead as important empirical phenomena to be investigated" (p. 2). This perspective does not argue for the "differences" as necessarily reflective of the group's adaptive solutions and coping. It clearly avoids as well the assumptions with the deficit model.

[8]There have also appeared a number of review articles, which critique the above models, and characterize the related literature into various conceptual frameworks (similar to what we are doing in this review). Though the content would be too lengthy to discuss here, we would like to alert interested readers to a number of these writings. They are: McCarthy and Yancey, 1971; Johnson, 1974; Nobles, 1974; Staples, 1974; Gurin and Epps, 1975; Braxton, 1976; Dennis, 1976; Allen, 1978; and Mathis, 1978.

white mind is capable of. What appears as differences are the results of experience and training. When used *correctly*, existing psychological principles are applicable to all" (p. 23).

This viewpoint is one of very few which considers a general psychology as applicable to all. Instead, many writings express an urgency for the creation of a separate Black Psychology (for example, Clark, 1972; Nobles, 1972; White, 1972; Gordon, 1973). Hayes (1972) rejects the idea of creating a Black Psychology, American psychology already being extremely varied in assumptions, methodology, and even definitions of behavior. However, he does suggest "radical black behaviorism." Based on the premises of radical behaviorism offered by Skinner (1964) and Day (1969), and the experimental analysis of behavior "upon which a radically objective science of human behavior can be developed," Hayes offers this as an alternative to the more diffuse black psychology proposed by others. "As a potential science, it rejects the mentalistic concepts of contemporary psychology, is individualistic in its approach, and has immediate utility for meeting the needs of black people" (p. 58).

In one of the most exteme positions, Mosby (1972) sees racially different people as warranting separate scientific study. Furthermore, "the differences lead to the observation that conceptualizations of the black personality should be based separately on black males and black females. Also, separate norms, within sex divisions, for different locales and cultural situations (Northern versus Southern) should exist. If we are to avoid generalizations that would place all blacks in neurotic and psychotic categories, individual personality assessment and measures for the adequacy of that assessment should be based on norms from groups comparable in sex, race, and culture" (p. 41).

These last two examples point to where a possible distinction in the arguments might be drawn. What is really meant by a "Black Psychology?" There is no question that there have been serious racial "misuses" in the psychological realm, and that in both research and practice (testing and therapy), attention should be paid to racial differences. However, that is a different issue than the theoretical question of whether explanations of human behavior should be based on differenct definitions of structure and organization, according to race. Most of the people calling for a "Black Psychology" seem to be focusing more on the problems of practice, of application. An exception to this is Nobles (1972), who gives an historical account of how philosophical assumptions have played a large part in shaping psychology; and suggests the need for a "Black Psychology" which is based on the philosophical assumptions of black people (their philosophical assumptions being different in some ways from the assumptions of whites). Another example of the focus being more on the theoretical aspect is the argument already mentioned, made by Comer and Poussaint, who view a general psychology as applicable to all people acquiring different patterns in personality organization as a result of their unique experience.[9]

CONCLUSIONS

As we have argued within most of the previous chapters, identity formation is a complex developmental process influenced by a wide range of dimensions, which include intrapsychic as well as sociocultural forces. What our study examined was not only whether members of different racial groups have different "amounts" (high or low) of self-esteem, or whether self-concept constructions vary across racial lines. We also addressed questions such as what features in the lives of adolescents shape and influence these self-image appraisals. Many factors, originally interpreted in terms of our Q-sort data and clinical interview material, have also been given varying degrees of significance by the empirical and theoretical writings which appeared since the initial publication of this study. While the unique characteristics of these many studies do not provide an opportunity for direct support (or contradiction) of our original interpretations, we have highlighted a number of important issues to be considered in this area of research.

There are at least two significant sets of changes which have occurred since initial publication of our findings. These concern situations (social, political, economic) and interpretations. Clearly, major societal constraints have been modified. In some areas (such as professional training) these modifications have probably contributed to some increased opportunities for black adolescents. Admittedly, this is an observation made from outside of the black community. It would be important to understand the actual extent of these changes as well as how black adolescents experience these changes before jumping to the conclusion that "blacks" are no longer subject to social forces markedly different from those encountered by "whites."

In terms of interpretations, we have shown how a particular perspective may shape the entire research endeavor; and how explanatory models through their prominence in the field over a given period of time, lead to generally prescribed, plausible interpretations. More

---

[9]While the enthusiasm for the creation of a "Black Psychology" ensues, there are others who do not even consider race as a relevant variable for personality research or theory (Edwards, 1974), or barely significant as a predictor of general psychopathology (Warheit, *et al.*, 1975).

specifically, the most important characteristic of the literature of the past decade concerns a shift in perspective, from the deficit model to the adaptation/functional model. Our original study can clearly be viewed through the lenses offered by the functional perspective, leading us to now ask about the adaptive value of "identity foreclosure," rather than view this pattern as evidence of "pathology" and of the destructiveness of white-based social institutions.

Self-Image Significance Tables

For the purpose of complete recording of all comparative correlation data, several cutting points are used in reporting the following results of nonparametric tests:

1. All differences at or below .05 probability are given at their exact probability levels. These differences are clearly *significant* ones.

2. All differences which are at probability levels between .05 and .20 are also reported at their exact probability levels. These differences represent *trends* in the findings.

3. All differences which are at a probability level above .20 are reported as >.20. These results suggest, at most, *tendencies* in the data, and are not treated as interpretable findings in any more specific manner than this.

4. Where no differences at any probability level were found, the result is reported as "B = W."

In interpreting the data[1] it is obviously the results occurring at or below .05 probability which are most important and most relied upon. However, for the sake of completeness and appreciation of tendencies and trend patterns, the other two cutting points are also included.

---

[1] The complete matrices of correlations from all subjects can be obtained by contacting the senior author.

Table 1. Interracial Comparisons of Parental Self-Image Intrayear Pearson Correlations

Year	N^a	me/pa Comparison	P^b	me/ma Comparison	P	me/ma Comparison	P
1	10 B; 10 W	B > W	.05	B > W	.10	B = W	—
2	7 B; 6 W	B > W	.04	B > W	.03	B > W	.10
3	7 B; 5 W	B > W	.10	B > W	.13	B > W	> .20
4	7 B; 4 W	B > W	.13	B > W	> .20	W > B	.14

[a] The symbol "B" represents black subjects; the symbol "W" represents white subjects.
[b] As determined by the Mann-Whitney U Test.

Table 2. Interracial Comparisons of Parental Self-Images: Interyear Pearson Correlations

Interval	N	pa Comparison	P^a	ma Comparison	P
1-2	7 B; 6 W	B > W	> .20	B > W	.18
1-3	7 B; 6 W	B > W	> .20	B > W	> .20
1-4	7 B; 4 W	B > W	.16	B > W	> .20
2-3	5 B; 5 W	B > W	.13	B > W	.11
2-4	5 B; 4 W	B > W	.03	B > W	.11
3-4	7 B; 4 W	W > B	> .20	B > W	> .20

[a] As determined by the Mann-Whitney U Test.

Table 3. Interracial comparisons of Personal Time: Self-Image Intrayear Pearson Correlations

Year	N	Comparison	P[a]	me/future Comparison	P	past/future Comparison	P	future average Comparison	P
1	10 B; 10 W	B=W	—	B=W	—	B=W	—	B>W	.05
2	7 B; 6 W	B>W	>.20	B>W	.13	W>B	.20	B=W	—
3	7 B; 5 W	B=W	—	B>W	.17	B=W	—	B>W	>.20
4	7 B; 4 W	B=W	—	B=W	—	W>B	.20	B>W	>.20

[a] As determined by the Mann-Whitney U Test.

Table 4. *Changes* of Black Subjects: Intrayear Pearson Correlations for Personal Time Self-Images

Year Interval	N	me/past Comparison	Pa	me/future Comparison	P	past/future Comparison	P	future average Comparison	P
1-2	7 B; 6 W	Same[b]	—	Same	—	Lower	.20	Same	—
1-3	7 B; 6 W	Same	—	Higher	.05	Same	—	Same	—
1-4	7 B; 4 W	Lower	.20	Higher	.05	Same	—	Same	—
2-3	5 B; 5 W	Same	—	Same	—	Higher	.20	Lower	.20
2-4	5 B; 4 W	Lower	.20	Higher	.06	Lower	.20	Lower	.20
3-4	7 B; 4 W	Lower	.20	Same	—	Lower	.20	Lower	.20

[a] As determined by Wilcoxon and sign tests.
[b] Entry of "same," "lower," or "higher" refers to comparison of later year with earlier year.

Table 5. *Changes* of White Subjects' Intrayear Pearson Correlations for Personal Time Self-Images

Year Interval	N	me/past		me/future		past/future		past average	
		Comparison[b]	P	Comparison	P	Comparison	P^a	Comparison	P
1-2	7 B; 6 W	Same	—	Same	—	Same	—	Same	—
1-3	7 B; 6 W	Lower	>.20	Higher	>.20	Lower	.06	Lower	>.20
1-4	7 B; 4 W	Same	—	Higher	4/4[c]	Same	—	Lower	4/4
2-3	5 B; 5 W	Same	—	Same	—	Lower	>.20	Lower	>.20
2-4	5 B; 4 W	Same	—	Higher	4/4	Same	—	Lower	4/4
3-4	7 B; 4 W	Same	—	Higher	4/4	Same	—	Same	—

[a] As determined by Wilcoxon and sign tests.
[b] Entry of "same," "lower," or "higher" refers to comparison of later year with earlier year.
[c] Result reported as such because specific probability level unavailable for this size N.

Table 6. Interracial Comparisons of Personal Time Self-Images: Interyear Pearson Correlations

Year Interval	N	past		future	
		Comparison	P^a	Comparison	P
1-2	7 B; 6 W	B > W	.08	B > W	.08
1-3	7 B; 6 W	B > W	> .20	B > W	> .20
1-4	7 B; 4 W	B > W	> .20	B > W	.13
2-3	5 B; 5 W	B > W	> .20	B > W	.13
2-4	5 B; 4 W	B > W	> .20	B > W	> .20
3-4	7 B; 4 W	W > B	> .20	B = W	—

^{a} As determined by the Mann-Whitney U Test.

Table 7. Interracial Comparisons of *fantasy* Self-Image: Interyear Pearson Correlations

Year Interval	N	fantasy	
		Comparison	P^a
1-2	7 B; 6 W	B > W	.10
1-3	7 B; 6 W	B = W	—
1-4	7 B; 4 W	B > W	.16
2-3	5 B; 5 W	B > W	> .20
2-4	5 B; 4 W	B > W	.04
1-4	7 B; 4 W	B > W	.16

^{a} As determined by the Mann-Whitney U Test.

Table 8. Interracial Comparisons of Current Self-Image (*me*): Average Intrayear Pearson Correlations

Year	N	Average me^a	
		Comparison	P^a
1	10 B; 10 W	B = W	—
2	7 B; 6 W	B > W	.16
3	7 B; 5 W	B > W	> .20
4	7 B; 4 W	B = W	—

^{a} The average *me* is calculated using Z conversions of the seven other self-images with which it occurs in each of the years; for example, *me/ma*, *me/friend*, and so on.
^{b} As determined by the Mann-Whitney U Test.

Table 9. Interracial Comparisons of Current Self-Image (*me*): Interyear Pearson Correlations

		me	
Year Interval	N	Comparison	P
1-2	7 B; 6 W	B = W	—
1-3	7 B; 6 W	B > W	> .20
1-4	7 B; 4 W	B > W	> .20
2-3	5 B; 5 W	B > W	.03
2-4	5 B; 4 W	B > W	.12
3-4	7 B; 4 W	B > W	.10

As determined by the Mann-Whitney U Test.

Table 10. Interracial Comparisons of Individual Subject's *Changes* in Averages of Intrayear Pearson Correlations (*Structural Integration*)[a]

Year Interval	N	Black	P[b]	White	P
1-2	7 B; 6 W	Same	—	Same	—
1-3	7 B; 6 W	Higher	.12	Lower	> .20
1-4	7 B; 4 W	Same	—	Higher	3/4[c]
2-3	5 B; 5 W	Higher	> .20	Same	—
2-4	5 B; 4 W	Same	—	Higher	3/4
3-4	7 B; 4 W	Same	—	Higher	4/4

[a] This table shows the blacks with tendencies towards increasing values in intervals Years 1-3 and 2-3. Such a difference does not appear in Figure 16. The apparent discrepancy is attributable to the fact that the table is calculated from *matched-pair* comparisons (same subject), whereas the figure represents group tendencies (medians of the entire group) and thus even includes some values not included in the table. Despite this difference in the two displays of data, the table also again demonstrates that the whites are the subjects who are the "changers."
[b] As determined by the sign test using comparisons of averages for same subjects.
[c] Where fractions are given, they represent the *actual* number who change in the stated direction, as compared to the total number of subjects in that particular interval.

Table 11. Interracial Comparisons of Interyear Differences in Yearly Averages of Intrayear Pearson Correlations (*Structural Integration*)

Year Interval	N	Black Intrayear Averages	White Intrayear Averages	P^a
1-2	7 B; 6 W	$B_2 = B_1{}^b$	$W_2 = W_1$	—
1-3	7 B; 6 W	$B_3 = B_1$	$W_3 > W_1$	.06
1-4	7 B; 4 W	$B_4 = B_1$	$W_4 > W_1$	.20
2-3	5 B; 5 W	$B_3 = B_2$	$W_3 = W_2$	—
2-4	5 B; 4 W	$B_4 = B_2$	$W_4 > W_2$	>.20
3-4	7 B; 4 W	$B_4 = B_3$	$W_4 = W_3$	—

[a] As determined by the Mann-Whitney U Test. Although the years have, in part, the same Ss, several years have different numbers of Ss. To use a related samples test would therefore have meant excluding several results. Hence the Mann-Whitney test is again applied here.
[b] Numerical subscript refers to the corresponding year.

Table 12. Interracial Comparisons of Average Interyear Pearson Correlations (*Temporal Stability*)

Year Interval	N	Interyear Averages	P^a
1-2	7 B; 6 W	B>W	.04
1-3	7 B; 6 W	B>W	.12
1-4	7 B; 4 W	B>W	.20
2-3	5 B; 5 W	B>W	.20
2-4	5 B; 4 W	B>W	.04
3-4	7 B; 4 W	B=W	—

[a] As determined by the Mann-Whitney U Test.

Table 13. Comparisons of Blacks with Fathers versus Blacks without Fathers for the Magnitude of Their *pa* Interyear Correlation

		pa	
Year Interval	N	Comparison	P^a
1-2	7	Bo>Bp[b]	.15
1-3	7	Bo>Bp	.03
1-4	7	Bo>Bp	.20
2-3	7	Bo>Bp	.10
2-4	7	Bo>Bp	.20
3-4	7	Bo>Bp	.20

[a] As determined by the Mann-Whitney U Test.
[b] Bo=subjects without fathers. Bp=subjects with fathers.

Table 14. Comparisons of Blacks with Fathers versus Blacks without Fathers for Magnitude of Their *me/pa* Correlation

		me/pa	
Year	N	Comparison	P^a
1	10	Bo>Bp[b]	.09
2	7	Bo>Bp	.06
3	7	Bo>Bp	.06
4	7	Bo=Bp	—

[a] As determined by the Mann-Whitney U Test.
[b] Bo=subjects without fathers; Bp=subjects with fathers.

Further Aspects of the Sample

Table 1. Additional Characteristics of Participants and Drop-outs.[a]

Subject	Race[b]	Geographic Mobility[c]	School Performance[d]	School Dropout	Part-Time Work	Number of Siblings
Dropouts						
JE	B	—	Average-above average	—	—	2
KB	B	—	Average-above average	—	—	1
DM	W	—	Below average	+	+	3
AL	W	+	Average-above average	—	+	3
ND	W	+	Below average	—	+	3
IQ	W	+	Failing	+	+	8
LE	W	—	Below average	?	+	0
Active Participants						
FM	B	+	Below average	—	+	5
GR	B	+	Average	—	+	5
LT	B	—	Below average	—	+	2
OS	B	—	Below average-failing	—	+	4
FL	B	—	Low average	+[e]	+	4
MD	B	+	Below average	—	+	1
TM	B	—	Average	—	+	4
BA	B	+	Low average	+	+	2
KE	B	—	Average	—	+	3
HN	PR	—	Average	—	+	5
NS	W	+	Low average	—	+	5
BT	W	+	Below average	—	+	6
VR	W	+	Average-above average	+[f]	+	1
JK	W	+	Average	—	+	2
JR	W	+	Average	—	+	2

[a] Other data, such as age and socioeconomic class are given in Table 1, Chapter 2.
[b] B, black; W, white; PR, Puerto Rican.
[c] Absence of a characteristic; presence of characteristic.
[d] As summarized from school records of the subject's last year of school attendance.
[e] Forced to leave school as a result of juvenile delinquency and subsequent reformatory sentence.
[f] Two-weeks duration.

Symbolic Formulations of Identity Variants

Listed in this section are "translations" of the operational definitions given in Chapter 3. The equations included here are abbreviations of complex processes. They are presented not to suggest that these processes can be simplified to one-line equations. The point rather is to suggest that there may be other forms of notation with which one can characterize these variants as empirical definitions, and understandings of the relevant variables are further elaborated.

Preceding each symbolic formulation is a summary of the variant's definition.

1. *Ego Identity Development.* Throughout childhood and adolescence, an ego identity develops, with an acceleration of this process occurring in adolescence. Less specific remarks have been directed toward the period after adolescence. It is implied, however that the process slowly continues, reaching a climax in old age, the "wisdom" stage of the life cycle (Erikson, 1960). We may describe an individual as manifesting identity formation when both the structural integration and the temporal stability of his self-images are simultaneously increasing over any given period of time, though not necessarily at the same rate.

Symbolically: $S * T = F$, where S = structural integration and may only increase; T = temporal stability and may only increase; F = a dependent variable, which may only increase; $*$ = an interaction between S and T.

2. *Identity Diffusion.* This clinical type designates those cases in which there is failure to achieve integration and continuity of self-image. The category is a broad one and probably includes several subtypes, for there are conceivably multiple etiologies underlying this outcome of identity formation. Such a state may be present at any stage of the life cycle. However, it is theoretically most manifest at adolescence when it hinders psychosocial development. Again, only through further empirical investigaton can this hypothesis of stage specificity be studied. For example, does identity diffusion occur during latency, or unexpectedly in adulthood, and lead to analogous complications in further psychosocial development?

Operationally, a person is said to be in a state of identity diffusion when both features of his self-images show a repeated decrease over any period of time. Again, these changes need not be occurring at the same rate. In fact, differing rates may be one sign of variant kinds of identity diffusion. However, of the two features, structural integration is the most critical here in defining the state of diffusion. A third possibility in this category is a form of "attenuated diffusion," in which either (a) structural integration decreases and temporal stability remains constant; or (b) the milder type, in which temporal stability decreases and structural integration remains constant. It is possible that these represent early forms of flagrant identity diffusion, where reversibility is more likely.

Symbolically: $S * T = D$, where S and T are as previously defined and now may only decrease over time; D = a dependent variable which may only decrease over time; * = an interaction between S and T.

Attenuated forms of diffusion are: Variant (a): $S * T = D$, where S must decrease over time; $T = K$ = constant; D = a dependent variable which decrease over time; * = as defined above. Variant (b): $S * T = D$, where T must decrease over time; $S = K$ = constant; D = a dependent variable which must decrease over time; * = as defined above.

3. *Identity Foreclosure.* This state superficially resembles identity development. There is a sense of integration, "purpose," stability, and a diminution in subjective confusion about these matters. However, the stability and purpose are reflections of an avoidance of alternatives, of a certain restrictiveness which eliminates any ambiguities. What appears to be the outcome of a successful process of identity formation is actually an impoverished, limited self-definition and sense of continuity. Operationally, a person is said to manifest identity foreclosure when either the structural integration and temporal stability of his self-images remain stable, or only temporal stability shows continued increase while structural integration is unchanging (See Chapter 3, footnote 7).

Symbolically: $S = T = K$, where K is a constant; S and T are defined as above; $S * T = C$, where C is a dependent variable which may only increase; K is a constant; S and T are defined as above; and $S = K$; * = interaction between S and T.

4. *Psychosocial Moratorium.* An individual is described as being in this state when he is "finding himself," experimenting with varied roles, new self-images, and future plans, at all costs remaining uncommitted to any particular alternative or identity. At the same time he is *not* tending toward, or in, a type of diffusion. This is, of course, the antithesis of foreclosure. Rather than rigid sameness, the content and types of self-images show continual variation. The key concept here is "openness," noncommitment. No irreversible decisions or plans are undertaken. A partly conscious, partly unconscious attempt is made to insure maximum flexibility and diversity before the further elimination of any possibilities, alternatives, or actions inherent in all stages of identity formation. Hypotheses as to when this stage occurs, and for whom, have been offered by Erikson.

Operationally, a person is in the period of psychosocial moratorium when the temporal stability of his self-images shows significant fluctuations (increasing and decreasing) over a given period of time. Consistent with the tendency running counter to diffusion, the structural integration feature shows less fluctuation than that of temporal stability, particularly in terms of any decrease in value.

Symbolically: (a) When the two functions change but in opposite directions over time, thus "compensating" for each other: $M = S * T$, where S and T are as defined above; $M = K =$ constant; $* =$ as defined above. (b) $S * T =$, where S is defined as above; $S = K$; $K =$ a constant; T is as defined above, and must alternatively increase and decrease over time; $M =$ a dependent variable. (c) $S * T = M$, where S is as defined above; T is as defined above; and both must independently increase of decrease over time; with any decrease over time; with any decrease in S being limited to *less than* $\frac{1}{2}$ $\Delta\ T$. $M =$ a dependent variable.

Examples of Q-Sort Statements

Table 1. Black Subjects: *Me* Self-image*

Rated high ("most important")
 I have heart
 I look at myself before I call someone a creep
 I think we should all go back to Jamaica
 I'm glad of my complexion
 I like to travel
 I build models
 I want to be a good doctor
 I want to be ambitious
 I like someone who helps
 I want to be an electrician
 I want enough luck to get a job
 I wish I were working now

Rated low ("least important")
 I'm called the "mad archer"
 I'm strong
 I break windows
 I think I'll buy a gas station
 I have 15 girl friends
 I think the biggest thing kids talk about is clothes
 I always wanted to be a scientist
 I'm the most popular guy
 I like to wrestle
 I like to do better

a Selected from three subjects over several years.

Table 2. White Subjects: *Me* Self-image[a]

Rated high ("most important")
    I score a lot in basketball
    I want to go to prison rather than to Russia
    I want to go into a good business
    I know everything going on in my girl friend
    I fight for freedom
    I go along with styles
    I'm too young to get serious
    I like shorts
    I want the Communist world wiped out

Rated low ("least important")
    I like shoes with buckles on the side
    I'm against my wife working
    I know 400 coloreds
    I fight for civil rights
    I'm underweight
    I like doctors to help me
    I like to be like the Pope
    I like to be like John Wayne

[a] Selected from two subjects over several years.

Bibliography

Abend, S.H. (1974). Problems of identity: Theoretical and clinical application. *Psychoanalytic Quarterly*, **43**: 606-637.

Abrahams, R. (1964). *Deep Down in the Jungle*. Hatboro, PA: Folklore Assoc.

Ackerman, N.W. (1951). Social role and total personality. *American Journal of Orthopsychiatry*, **21**: 1-17.

Adams, W.A. (1950). The negro patient in psychiatric treatment. *American Journal of Orthopsychiatry*, **20**: 305-310.

Adelson, J. (1961). The adolescent personality. Paper read at the meeting of American Psychological Association, mimeo.

Aichorn, A. (1935). *Wayward Youth*. New York: Viking Press.

Allen, W.R. (1978). The search for applicable theories of black family life. *Journal of Marriage and the Family*, **40**: 117-129.

Antonovsky, A., and Lerner, M. (1959). Occupational aspirations of lower class negro and white youth. *Social Problems*, **7**: 132-138.

August, G.J. and Felker, D.W. (1977). Role of affective meaningfulness and self-concept in the verbal learning styles of white and black children. *Journal of Educational Psychology*, **69**:253-260.

Bachman, J.G., Kahn, R.L., Médnick, M.T., Davidson, T.N. and Johnston, L.D. (1967). *Youth in Transition*, Vol. 1, Ann Arbor, MI: Institute for Social Research.

Baittle, B. (1961). Psychiatric aspects of the development of a street corner group. *American Journal of Orthopsychiatry*, **31**: 703-712.

Baker, F. (1971). Measures of ego identity: A multitrait-multimethod validation. *Educational Psychology Measurements*, **31**:165-174.

Baldwin, J. (1953). *Go Tell It on the Mountain*. New York: Knopf.

Baldwin, J. (1961). *Nobody Knows My Name*. New York: Dial Press.

Baldwin, J. (1962). *Another Country*. New York: Dial Press.

Baldwin, J. (1963). *The Fire Next Time*. New York: Dial Press.

Banks, C.W. (1976). White preference in blacks: A paradigm in search of a phenomenon. *Psychological Bulletin*, **83**:1179-1186.

Banks, J.A. (1972). Racial concept and the black self-concept. In Banks, J.A. and Grambs, J.D. (Eds.), *Black Concept*. New York: McGraw-Hill.

Barber, B. (1961). Social-class differences in educational life-chances. *Teachers College Record*, **63**: 94-101.

Baughman, E.E. (1971). *Black Americans*. New York: Academic Press.

Beardslee, D., and O'Dowd, D. (1962). Students and the occupational world. In Sanford, N. (Ed.), *The American College*. New York: Wiley.

Beglis, J.F. and Sheikh, A. (1974). Development of self-concept in black and white children. *Journal of Negro Education*, 43:1-4-110.

Bellak, L. (1954). *The Thematic Apperception Test and the Chldren's Apperception Test in Clinical Use*. New York: Grune and Stratton.

Bernard, J. (Ed.) (1961). Teenage culture. *Annals of the American Academy of Political and Social Sciences*, 338:1-12.

Bernfeld, S. (1938). Types of adolescence. *Psychoanalytic Quarterly,* 7:243-253.

Bialer, I. (1961). Conceptualization of success and failure in mentally retarded and normal children. *Journal of Personality*, 29:303-320.

Billingsley, A. (1968). *Black Families in White America*. Englewood Cliffs: Prentice-Hall.

Blaine, G.B. and McArthur, C.C. (1961). *Emotional Problems of the Student*. New York: Appleton.

Blake, T. and Dennis, W. (1943). The development of stereotypes concerning the negro. *Journal of Abnormal and Social Psychology*, 38:525-531.

Blau, P.M. and Duncan, O.D. (1967). *The American Occupational Structure*. New York: Wiley.

Blos, P. (1962). *On Adolescence*. New York: Macmillan (Glencoe Press).

Bogardus, E.S. (1933). A social distance scale. *Sociology and Social Research*, 17:265-271.

Bone, R.A. (1958). *The Negro Novel in America*. New Haven: Yale University Press.

Bolling, J.L. (1974). The changing self-concept of black children—The black identity test. *Journal of the National Medical Association*, 66:28-31, 34.

Bordua, D.J. (1961). Delinquent subcultures. *Annals of the American Academy of Political and Social Sciences*, 338:119-136.

Bourne, E. (1978a). The state of research on ego identity: A review and appraisal. Part I. *Journal of Youth and Adolescence*, 7:223-257.

Bourne, E. (1978b). The state of research on ego identity: A review and appraisal. Part II. *Journal of Youth and Adolescence*, 7:371-392.

Boyanowsky, E.O. and Allen, V.L. (1973). Ingroup norms and self-identity as determinants of discriminatory behavior. *Journal of Personality and Social Psychology*, 25:408-418.

Brand, E.S., Ruiz, R.A. and Padilla, A.M. (1974). Ethnic identification and preference: A review. *Psychological Bulletin*, 81:860-890.

Braxton, E.T. (1976). Structuring the black family for survival and growth. *Perspectives in Psychiatric Care*, 14:165-173.

Brigham, J.C. (1973). Ethnic stereotypes and attitudes: A different mode of analysis. *Journal of Personality*, 41:206-223.

Brigham, J.C. (1974). Views of black and white children concerning the distribution of personality characteristics. *Journal of Personality,* 42:144-158.

Brody, E.B. (1961). Social conflict and schizophrenic behavior in young adult negro males. *Psychiatry*, 24:337-346.

Brody, E.B. (1963). Color and identity conflict in young boys. *Archives of Clinical Psychiatry*, **10**:354-360.

Brody, E. (Ed.) (1968). *Minority Group Adolescents in the United States*. Baltimore: Williams and Wilkins.

Brown, B.R. (1966). *The Assessment of Self-Concept Among Four Year Old Negro and White Children: A Comparative Study Using the Brown IDS Self-Concept Referent Test*. Institute for Developmental Studies.

Brown, C. (1965). *Manchild in the Promised Land*. New York: Macmillan.

Brown, M.C. (1955). Status of jobs and occupations as evaluated by an urban negro sample. *American Sociological Review*, **20**:561-566.

Brown, N. (1977). Personality characteristics of black adolescents. *Adolescence*, **12**:81-87.

Brown, S.R. (1968). Bibliography on Q technique and its methodology. *Perceptual and Motor Skills*, **26**:587-613.

Budyk, J. (1961). Female adolescent development in the lower class. Unpublished undergraduate paper, Radcliffe College.

Burbach, H.J. and Brideman, B. (1976). Relation between self-esteem and locus of control in black and white 5th graders. *Child Study Journal*, **6**:33-37.

Busk, P.L., Ford, R.C., and Shulman, J. (1973). Effect of schools' racial composition on the self-concept of black and white students. *Journal of Educational Research*, **67**:57-63.

Carey, P. and Allen, D. (1977). Black studies: Expectation and impact on self-esteem and academic performance. *Social Science Quarterly*, **57**:811-820.

Carter, J.H. (1972). The black struggle for identity. *Journal of the National Medical Association*, **64**:236-238.

Chein, I., Gerard, D.L., Lee, R.S. and Rosenfeld, E. (1964). *The Road to H: Narcotics, Delinquency and Social Policy*. New York: Basic Books.

Chestang, L. (1970). The issue of race in casework practice. *Social Work Practice*. New York: Columbia University Press.

Christmas, J.J. (1973). Self-concept and attitudes. In Miller, K.S. and Dreger, R.M. (Eds.) *Comparative Studies of Blacks and Whites in the United States*. New York: Seminar Press.

Cicirelli, V.G. (1977). Relationship of socioeconomic status and ethnicity to primary grade children's self-concept. *Psychology in the Schools*, **14**:213-215.

Clark, C. (1972). Black studies or the study of black people? In Jones, R.L. (Ed.), *Black Psychology*. New York: Harper and Row.

Clark, E.T. (1967). The Clark U-scale. (Unpublished test.) New York: St. John's University.

Clark, K. (Ed.) (1963). *The Negro Protest*. Boston: Beacon Press.

Clark, K. (1965). *Dark Ghetto: Dilemmas of Social Power*. New York: Harper.

Clark, K.B. and Clark, M.K. (1939). The development of consciousness of self and emergence of racial identification in Negro preschool children. *Journal of Social Psychology,* **10**:591-599.

Clark, K.B. and Clark, M.P. (1947). Racial identification and preference in Negro children. In Newcomb, T.M. and Hartley, E.L. (Eds.), *Readings in Social Psychology.* New York: Holt, Rinehart and Winston.

Cleaver, E. (1968). *Soul on Ice.* New York: Delta.

Cohen, A.K. (1955). *Delinquent Boys.* New York: Macmillan (Glencoe Press).

Coleman, J.S. (1961). *The Adolescent Society.* New York: Macmillan (Glencoe Press).

Coles, R. (1963). Southern children under desegregation. *American Journal of Psychiatry,* **120**:332-344.

Coles, R. (1964). A matter of territory. *Journal of Social Issues,* **20**:43-53.

Coles, R. (1965a). Private problems and public evil: Psychiatry and segregation. *Yale Review,* **14**:513-531.

Coles, R. (1965b). Racial conflict and a child's question. *Journal of Nervous and Mental Disease,* **140**:162-170.

Coles, R. (1965c). It's the same but it's different. *Daedalus,* **94**:1107-1132.

Coles, R. (1967). *Children of Crisis.* Boston: Little Brown.

Colley, T. (1959). Psychological sexual identity. *Psychological Review,* **66**:165-177.

Comer, J.P. and Poussaint, A.V. (1965). *Black Child Care.* New York: Simon & Schuster.

Cooley, C.H. (1902). *Human Nature and the Social Order.* New York: Scribner's.

Coopersmith, S. (1967). *The Antecedents of Self-Esteem.* San Francisco: W.H. Freeman and Co.

Crain, R.L. and Weisman, C.S. (1972). *Discrimination, Personality, and Achievement: A Survey of Northern Blacks.* New York: Seminar Press.

Crowne, D.P. and Marlowe, D. (1964). *The Approval Motive: Studies in Evaluative Dependence.* New York: Wiley.

Cummings, E.E. (1969). *I: Six Nonlectures.* Cambridge: Harvard University Press.

Curran, F.J., and Frosh, J. (1942). The body image in adolescent boys. *Journal of Genetic Psychology,* **60**:37-60.

Dahlstrom, W.G. and Welsh, G.S. (1960). *An MMPI Handbook.* Minneapolis: University of Minnesota Press.

Damon, W. (1979). *The Social World of the Child.* San Francisco: Jossey-Bass.

D'Andrade, R.G. (1973). Father absence, identification, and identity. *Ethos,* **1**:440-445.

Darden, B.J. (1977). Self-concept and blacks' assessment of black leading roles in motion pictures and television. *Journal of Applied Psychology,* **62**:620-623.

Davids, A. (1973). Self-concept and mother concept in black and white preschool children. *Child Psychiatry and Human Development,* **4**:30-43.

Davids, A. and Lawton, M.J. (1961). Self-concept, mother-concept, and food aversions in emotionally disturbed and normal children. *Journal of Abnormal and Social Psychiatry*, **62**:309-314.

Dai, B. (1955). Some problems of personality development among negro children. In Kluckhohn, Clyde, et al. (Eds.) *Personality in Nature, Society and Culture*. New York: Knopf.

Davie, M.R. (1949). *Negroes in American Society*. New York: McGraw-Hill.

Davis, A., and Dollard, J. (1941). *Children of Bondage*. Washington: American Council on Education.

Davis, A. and Havighurst, R.J. (1946). Social class and color difference in child rearing. *American Sociological Review*, **11**:698-711.

Davis, K. (1958). Mental hygiene and class structure. In Stein, H., and Cloward, R. (Eds.) *Social Perspectives on Behavior*. New York: Macmillan (Glencoe Press).

DeLevita, D. (1976). On the PSA concept of identity. *International Journal of Psychoanalysis*, **47**:2-

Dennis, N. (1960). *Cards of Identity*. New York: Meridian.

Dennis, R.M. (1976). Theories of the black family: The weak-family and strong-family schools as competing ideologies. *Journal of Afro-American Issues*, **4**:315-328.

Denizen, N. (1966). The significant others of a college population. *Sociology Quarterly* **7**:315-328.

Derbyshire, R.L., and Brody, E. (1964a). Marginality, identity and behavior in the American negro: A functional analysis. *International Journal of Social Psychiatry*, **10**:7-13.

Derbyshire, R.L., and Brody, E. (1964b). Social distance and identity conflict in negro college students. *Sociology and Social Research*, **48**:301-314.

Deutsch, H. (1967). *Selected Problems of Adolescence*. New York: International Universities Press.

Dignan, M.H. (1965). Ego identity and maternal identification. *Journal of Personality and Social Psychology*, **1**:476-483.

Dinitz, S., Scarpitti, F.R., and Reckless, W.C. (1962). Delinquency vulnerability: A cross group and longitudinal analysis. *American Sociological Review*, **27**:515-517.

Dizard, J.E. (1970). Black identity, social class, and black power. *Psychiatry*, **33**:145-173.

Donovan, J.M. (1975). Identity status and interpersonal style. *Journal of Youth and Adolescence*, 4L37-55.

Dorris, R.J., Levinson, J., and Hanfmann, E. (1954). Authoritarian personality studied by a new variation of the sentence completion test. *Journal of Abnormal and Social Psychology*, **49**:99-108.

Dragstin, S.E. and Elder, G.H. Jr. (Eds.) (1975). *Adolescence in the Life Cycle*. New York: Wiley.

Drake, S.C., and Cayton, H.R. (1962). *Black Metropolis: A Study of Negro Life in a Northern City.* New York: Harper.

Dreger, R.M., and Miller, K.S. (1960). Comparative psychological studies of negroes and whites in the United States. *Psychological Bulletin,*57:361-402.

Dreger, R.M., and Miller, K.S. (1968). Comparative psychological studies of negroes and whites in the United States. *Psychological Bulletin,* 70:1.

Dubois, C. (1944). *The People of Alor.* Minneapolis: University of Minnesota Press.

Dubois, W.E. (1903). *The Souls of Black Folk: Essays and Sketches.* Chicago: McClurg.

Earl, L., and Lohmann, N. (1978). Absent fathers and black male children. *Social Work,* 50:413-415.

Edwards, A.M. (1934). *Comparative Occupational Statistics for the United States.* U.S. Government Printing Office.

Edwards, D.W. (1971). The development of a questionnaire method of measuring exploration preferences. In Feldman, M.J. (Ed.) *Buffalo: Studies in Psychotherapy and Behavioral Change.* No. 2. *Theory and Research in Community Mental Health.* Buffalo: S.U.N.Y. at Buffalo.

Edwards, D.W. (1974). Black versus whites: When is race a relevant variable? *Journal of Personality and Social Psychiatry,* 29:39-49.

Elkin, F., and Westley, W.A. (1955). The myth of adolescent culture. *American Sociological Review,* 20:680-684.

Elkins, S.M. (1963). *Slavery.* New York: Grosset and Dunlap.

Ellison, R. (1952). *Invisible Man.* New York: Random House.

Encounter Editors. (1963). Negro crisis. *Encounter,* 21:August.

Epstein, R. (1963). Social class membership and early childhood memories. *Child Development,* 34:503-508.

Erikson, E. (1950). *Childhood and Society.* New York: Norton.

Erikson, E. (1956). Ego identity and the psycho-social moratorium. In Witner, H., et al. (Eds.), *New Perspectives on Delinquency.* Washington, D.C.: U. S. Government Printing Office.

Erikson, E., and Erikson, K. (1957). The conformation of the delinquent. *Chicago Review,* 10:15-23.

Erikson, E. (1958). *Young Man Luther.* New York: Norton.

Erikson, E. (1959). Identity and the life cycle. *Psychological Issues,* 1:1-171.

Erikson, E. (1960). The roots of virtue. In Huxley, J. (Ed.), *The Humanist Frame,* New York: Harper.

Erikson, E. (1962). Youth: Fidelity and diversity. *Daedalus,* 91:5-26.

Erikson, E. (1964). Memorandum on identity and negro youth. *Journal of Social Issues,* 20:29-42.

Erikson, E. (1968). *Identity: Youth and Crisis,* New York: Norton.

Erikson, E. (1969). *Ghandi's Truth.* New York: Norton.

Erikson, E.E. (1966). The concept of identity in race relations: Notes and queries. In Parsons, T., and Clark, K.B. (Eds.) *The Negro American.* Boston: Houghton-Mifflin.

Evans, J. (1950). *Three Men.* New York: Knopf.

Fanon, F. (1962). *Black Skin, White Masks.* New York: Grove Press.

Fisher, S., et al. (1957). Body boundaries and style of life. *Journal of Abnormal and Social Psychology.* 52:373-379.

Franck, K. and Rosen, E. (1949). A projective test of masculinity-femininity. *Journal of Consulting Psychiatry,* 13:247-256.

Frazier, E.F. (1939). *The Negro Family in the United States.* Chicago: University of Chicago Press.

Frazier, E.F. (1957). *Black Bourgeoisie.* New York: Macmillan (Glencoe Press).

Freedomways Editors. (1965). Freedomways: A quarterly review of the negro freedom movement. *Freedomways, 5.*

Freidenberg, E.Z. (1960). *The Vanishing Adolescent.* Boston: Beacon Press.

Freud, A. (1948). *The Ego and the Mechanisms of Defense.* London: Hogarth Press.

Fried, M., and Lindeman, E. (1961). Socio-cultural factors in mental health. *American Journal of Orthopsychiatry,* 31:87-101.

Fromm, E. (1956). *The Art of Loving.* New York: Harper.

Garcia, C., and Levenson, H. (1975). Differences between blacks' and whites' expectations of control by chance and powerful others. *Psychological Reports,* 37: 563-566.

Gestinger, S.H., Kunce, J.T., Miller, D.G., and Weinberg, S.R. (1972). Self-esteem measures and cultural disadvantagement. *Journal of Consulting and Clinical Psychology,* 38:149.

Gibbs, J.T. (1974). Patterns of adaptation among black students at a predominantly white university: Selected case studies. *American Journal of Orthopsychiatry,* 44:728-740.

Ginzburg, E., et al. (1951). *Occupational Choice.* New York: Columbia University Press.

Gladstone, H.P. (1962). Psychotherapeutic techniques with youthful offenders. *Psychiatry,* 147-159.

Glaser, D. (1958). Dynamics of ethnic identification. *American Sociological Review,* 23:31-40.

Glenn, N.D. (1963). Negro prestige criteria. *American Journal of Sociology,* 68:645-657.

Glicksberg, G. (1960). Psychoanalysis and the negro. *Phylon,* 21:337.

Goffman, E. (1963). *Stigma.* Englewood Cliffs, New Jersey: Prentice-Hall.

Goldschmid, M.L. (1970). *Black Americans and White Racism.* New York: Holt, Rinehart and Winston.

Goodman, A.J. (1972). Institutional racism: The crucible of black identity. In Banks, J.A. and Grambs, J.D. (Eds.) *Black Self-Concept.* New York: McGraw Hill.

Goodman, P. (1960). *Growing Up Absurd.* New York: Random House.

Gordon, T. (1973). Notes on white and black psychology. *Journal of Social Issues,* 29:87-96.

Gordon, V.V. (1976). The methodologies of black self-concept research: A critique. *Journal of Afro-American Issues,* 4:373-381.

Graubard, S.R. (1962). Youth: Change and challenge. *Daedalus,* 91:1-239.

Greenacre, P. (1958). Early physical determinants in the development of a sense of identity. *Journal of the American Psychoanalytic Association*, 6:612-627.

Grier, W.H., and Cobb, P.M. (1968). *Black Rage*. New York: Basic Books.

Griffin, J.H. (1961). *Black Like Me*. Boston: Houghton Mifflin.

Gurin, P. and Epps, E.G. (1975). *Black Consciousness, Identity, and Achievement*. New York: Wiley.

Haan, N. (1965). Coping and defense mechanisms related to personality inventories. *Journal of Consulting Psychiatry*, 29:373-378.

Hall, W.S., Cross, W.E., and Freedle, R. (1974). Stages in the development of black awareness: An exploratory investigation. In Jones, R.L. (Ed.), *Black Psychology*. New York: Harper and Row.

Hannerz, U. (1969). *Soulside*. New York: Columbia University Press.

Hansberry, L. (1959). *A Raisin in the Sun*. New York: Random House.

Hartmann, H. (1939). *Ego Psychology and the Problem of Adaptation*. New York: International Universities Press.

Hartmann, H., and Kris, E. (1945). The genetic approach in psychoanalysis. *The Psychoanalytic Study of the Child*, 1:11-30, New York: International Universities Press.

Hartmann, H., Kris, E., and Lowenstein, R.M. (1949). Some psychoanalytic comments on culture and personality. In Wilbur, G. (Ed.), *Psychoanalysis and Culture*. New York: International Universities Press.

Hartnagel, T.F. (1970). Father absence and self conception among lower class white and negro boys. *Social Problems*, 18:152-163.

Hauser, S.T. (1962). Patterns of estrangement. Unpublished manuscript, Harvard University.

Hauser, S.T. (1966). *Racial and Social Contexts of Ego Identity*. M.D. Thesis, Yale University School of Medicine.

Hauser, S.T. (1972). Adolescent self-image development: Longitudinal studies of black and white boys. *Archives of General Psychiatry*, 27:537-541.

Hauser, S. (1973). Thinking about race and racism: Clinical and Research Dilemmas. *Int. Journal of Group Psychotherapy*, 23: 242-259.

Hauser, S.T. (1976). Fantasy Disclosure, Interpersonal Behavior and Ego Development in Interracial Interviews. Ph.D Dissertation. Cambridge: Harvard University.

Hauser, S., Beardslee, W., Jacobson, A., and Noam, G. Longitudinal Studies of Ego Defenses and Ego Development in Early Adolescents, presented at Research in Progress Session, Fall Meeting, American Psychoanalytic Association, 1979.

Hauser, S.T. (1980). Familial Contexts of Ego Development and Self-Image Integration in Diabetic Adolescents: Longitudinal Studies, in *Proceedings of National Conference on Behavioral and Psychological Aspects of Diabetes*, Washington, D.C.: Government Printing Office, in press.

Havighurst, R.J., Bowman, P.H., Matthews, C.V., and Pierce, J.V. (1962). *Growing Up in River City*. New York: Wiley.

Hayes, W.H. (1972). Radical black behaviorism. In Jones, R.L. (Ed.) *Black Psychology*. New York: Harper and Row.

Healy, G.W. and DeBlassie, R.R. (1974). A comparison of Negro, Anglo, and Spanish-American adolescents' self-concepts. *Adolescence*, 9:15-24.

Hearn, R. (1963). Notes on negro life in a New Haven neighborhood. Unpublished mimeo.

Heiss, J., and Owens, S. (1972). Self-evaluations of blacks and whites. *American Journal of Sociology*, 78:360-370.

Hernton, C.C. (1965). *Sex and Racism in America*. Garden City: Doubleday.

Himes, J.S. Negro teenage culture. *Annals of the American Academy of Political and Social Science*, 338:91-101.

Hirsch, J.G. (1974). The emotions of social change: On cop-outs and rip-offs. *American Academy of Child Psychiatry*, 13:286-299.

Hollingshead, A. (1949). *Elmtown's Youth*. New York: Wiley.

Hollingshead, A. (1957). Two factor index of social class. New Haven: Yale University, mimeo.

Hollingshead, A., and Redlich, F.C. (1958). *Social Class and Mental Illness: A Community Study*. New York: Wiley.

Howard, L.P. (1960). Identity conflicts in adolescent girls. *Smith College Studies in Social Work*, 31:1-21.

Hraba, J., and Grant, G. (1970). Black is beautiful: A reexamination of racial preference and identification. *Journal of Personality and Social Psychiatry*, 16:398-402.

Hunt, L. and Hunt, J.G. (1975). Race and the father-son connection: The conditional relevance of father absence for the orientations and identities of adolescent boys. *Social Problems*, 23:35-52.

Hyman, H.H. (1959). Value systems of different classes. In Stein, H., and Cloward, R. (Eds.), *Social Perspectives on Behavior*. New York: Macmillan (Glencoe Press).

Isaacs, H.R. (1962). *The New World of Negro Americans*. New York: J. Day.

Jackson, D.N. (1967). *Personality Research Form*. New York: Research Psychologists Press.

Jackson, J.J. (1973). Family organization and technology. In Miller, K.S. and Dreger, R.M. (Eds.), *Comparative Studies of Blacks and Whites in the United States*. New York: Seminar Press.

Jacobson, E. (1964). *The Self and the Object World*. New York: International Universities Press.

Jegede, R.O. (1976). The identity status of Nigerian university students. *Journal of Social Psychology*, 100:175-179.

Johnson, E.E. (1973). Social perceptions and attitudes. In Miller, K.S. and Dreger, R.M. (Eds.), *Comparative Studies of Blacks and Whites in the United States*. New York: Seminar Press.

Johnson, L.B. (1974). Relevant literature in the study of the black family: An annotated bibliography. *Journal of Social and Behavioral Sciences*, 19:79-101.

Jones, D.L. (1979). African-American clients: Clinical practice issues. *Social Work*, March:112-118.

Jones, E. (1974). Social class and psychotherapy: A critical review of research. *Psychiatry*, **39**:307-320.

Jones, E.E. (1978). Black-white personality differences: Another look. *Journal of Personality Assessment*, **42**:244-252.

Jones, J. (1972). *Prejudice and Racism*. Reading, MA: Addison-Wesley.

Jones, L. (1964). *Dutchman and the Slave, Two Plays*. New York: Morrow.

Jones, L. (1976). *Tales*. New York: Grove Press.

Josselson, R.L., Greenberger, E., and McConochie, D. (1977a). Phenomenological aspects of psychosocial maturity in adolescence, Part I: Boys. *Journal of Youth and Adolescence*, **6**:25-55.

Josselson, R.L., Greenberger, E., and McConochie, D. (1977b). Phenomenological aspects of psychosocial maturity in adolescence, Part II: Girls. *Journal of Youth and Adolescence*, **6**:145-167.

Kaplan, B. (Ed.) (1961). *Studying Personality Cross-Culturally*. Evanston, IL: Row, Peterson.

Kardiner, A., and Ovessey, L. (1951). *The Mark of Oppression*. New York: Norton.

Karon, B.P. (1958). *The Negro Personality*. New York: Springer.

Katz, D., and Braly, K.W. (1933). Verbal stereotypes and racial prejudice. *Journal of Abnormal and Social Psychiatry*, **28**:280-290.

Katz, I. (1969). A critique of personality approaches to negro personality with research suggestions. *Journal of Social Issues*, **25**:13-27.

Katz, I., Cole, O.J., and Baron, R.M. (1976). Self-evaluation, social reinforcement, and academic achievement of black and white schoolchildren. *Child Development*, **47**:368-374.

Katzenmeyer, W.G., Stenner, A., Jackson. (1977). Estimation of the invariance of factor structures across sex and race with implications for hypothesis testing. *Educational and Psychological Measurement*, **37**:111-119.

Keil, C. (1966). *Urban Blues*. Chicago: University of Chicago Press.

Kellam, S.G., Ensminger, M.E. and Turner, R.J. (1977). Family structure and the mental health of children. *Archives of General Psychiatry*, **34**:1012.

Keller, S. (1963). The social world of the urban slum child: Some early findings. *American Journal of Orthopsychiatry*, **32**:823-831.

Kelly, J.G. (1969). Naturalistic observations in contrasting high schools. In Willems, E.P. and Rausch, H.L. (Eds.), *Naturalistic Viewpoints in Psychological Research*. New York: Holt, Rinehart and Winston.

Keniston, K., and Scott, P. (1959). *Exploratory research on identity*. Mimeo.

Keniston, K. (1961). Alienation and the decline of Utopia. *American Scholar*, **29**:161-200.

Keniston, K. (1962). Social change and youth in America. *Daedalus*, **91**:145-171.

Keniston, K. (1966). *The Uncommitted: Alienated Youth in American Society*. New York: Harcourt.

Kinder, D.R. and Reeder, L.G. (1975). Ethnic differences in beliefs about control. *Sociometry*, **38**:261-272.

King, J.R. (1976). African survivals in the black American family: Key factors in stability. *Journal of Afro-American Issues*, **4**:153-167.

Klapp, O. (1962). *Heroes, Fools and Villains*. Englewood Cliffs, New Jersey: Spectrum, Prentice-Hall.

Klausner, S.Z. (1964). Social class and self-concept. *Journal of Social Psychology*, **38**:201-205.

Kohn, M.L. (1963). Social class and parent-child relationships: An interpretation. *American Journal of Sociology*, **68**:471-480.

Korbin, S. (1961). Sociological aspects of the development of a street corner group. *American Journal of Orthopsychiatry*, **31**:685-702.

Korbin, S. (1962). Impact of cultural factors on the selected problems of adolescent development in the middle and lower class. *American Journal of Orthopsychiatry*, **32**:387-390.

Kovel, J. (1970). *White Racism*. New York: Pantheon.

Krate, R., Levanthal, G., Silverstein, B. (1974). Self-perceived transformation of negro-to-black identity. *Psychological Reports*, **35**:1071-1075.

Kuhn, M. (1964). The reference group reconsidered. *Sociological Quarterly*, **5**:6-12.

Langner, T. (1962). A twenty-two item screening score of psychiatric symptoms indicating impairment. *Journal of Health and Human Behavior*, **3**:269-276.

Lederer, W. (1964). Dragons, delinquents and destiny. *Psychological Monographs* #15., 1-80.

Leighton, A.H. (1959). *My Name is Legion*. New York: Basic Books.

Levenson, H. (1974). Activism and powerful others: Distinctions within the concept of internal-external control. *Journal of Personality Assessment*, **38**:377-383.

Levitt, M., and Rubenstein, B. (1964). Some observations on the relationship between cultured variants and emotional disorders. *American Journal of Orthopsychiatry*, **34**:423-432.

Lichenstein, H. (1961). Identity and sexuality: A study of their interrelationships in man. *Journal of the American Psychoanalytic Association*, **9**:179-206.

Liebow, E. (1967). *Tally's Corner*. Boston: Little, Brown.

Lifton, R.J. (1961). *Thought Reform and the Psychology of Totalism*. New York: Norton.

Lifton, R.J. (1962). Youth and history: Individual change in post war Japan. *Daedalus*, **91**:172-197.

Lincoln, C.E. (1962). *The Black Muslims in America*. Boston: Beacon Press.

Loevinger, J. (1976). *Ego Development*. San Francisco: Jossey-Bass.

Lomas, P. (1961). Family role and identity formation. *International Journal of Psychoanalysis*, **42**:371-380.

Lomax, L.E. (1962). *The Negro Revolt*. New York: Harper.

Lomax, L.E. (1963). *When the Word is Given*. New York: World.

Lott, A., and Lott, B.E. (1963). *Negro and White Youth: A Psychological Study in a Border State Community*. New York: Holt.

Lynd, H.M. (1958). *On Shame and the Search for Identity*. New York: Harcourt.

McCarthy, J.D. and Yancey, W.L. (1971). Uncle Tom and Mr. Charlie: Metaphysical pathos in the study of racism and personal disorganization. *American Journal of Sociology*, 76:648-652.

McDaniel, P.A., and Babchuk, N. (1960). Negro conceptions of white people in a northeastern city. *Phylon*, 21:7-19.

Machover, K. (1949). *Personality Projection in the Drawing of the Human Figure*. Springfield, Ill.: Charles C. Thomas.

MacInnes, C. (1963). Dark angel. *Encounter*, 21, August.

McNemar, Q. (1955). *Psychological Statistics*. New York: Wiley.

Maldonado, D. (1975). Ethnic self-identity and self-understanding. *Social Casework*, 56:618-622.

Malcolm X. (1968). *The Autobiography of Malcolm X*. New York: Grove Press.

Marcia, J.E. (1966). Development and validation of ego identity status. *Journal of Abnormal and Social Psychology*, 3:551-558.

Marcia, J.E. (1967). Ego identity status: Relationship to change in self-esteem, 'general maladjustment' and authoritarianism. *Journal of Personality*, 35:119-133.

Marcia, J.E. (1976). Identity six years after: A follow-up study. *Journal of Youth and Adolescence*, 5:145-160.

Marcia, J.E. (1978). Identity in adolescence. In Adelson, J. (Ed.). *Handbook of Adolescence*. New York: Wiley (in press).

Martin, D.G. (1969). Consistency of self-descriptions under different roles sets in neurotic and normal adults and adolescents. *Journal of Abnormal Psychology*, 74: 2, 173-176.

Martinek, T.J., and Cheffers, J.T.F., and Zaichowsky, L.D. (1978). Physical activity, motor development and self-concept: Race and age differences. *Perceptual and Motor Skills*, 46:147-154.

Mathis, A. (1978). Contrasting approaches to the study of black families. *Journal of Marriage and the Family*, 40:667-676.

Matteson, D.R. (1974). Alienation Versus Exploration and Commitment: Personality and Family Corollaries of Adolescent Identity Statuses. Report from the Project for Youth Research. Copenhagen: Royal Danish School of Educational Studies.

Mayovich, M.K. (1973). Stereotypes and racial images—white, black and yellow. *International Journal of Social Psychiatry*, 18:239-253.

Mead, G.H. (1934). *The Social Psychology of George Herbert Mead*. Strauss, A. (Ed.). Chicago: University of Chicago Press.

Milgram, N.A. (1971). Locus of control in Negro and white children at four age levels. *Psychology Reports*, 29:459-465.

Miller, W. (1959). *The Cool World*. New York: Little , Brown.

Miller, W.B. (1958). Lower class culture as a generating milieu of gang delinquency. *Journal of Social Issues*, **14**:5-19.

Miller, T.W. (1975). Effects of maternal age, education, and employment status on self-esteem of the child. *Journal of Social Psychiatry*, **95**:141-142.

Moore, T.L. (1975). Training of white adolescents to accurately simulate black adolescent personality. *Adolescence*, **10**:231-239.

Mosby, D.P. (1972). Toward a new specialty of black psychology. In Jones, R.L. (Ed.), *Black Psychology*. New York: Harper and Row.

Moynihan, D.P. (1965). Employment, income, and the ordeal of the negro family. *Daedalus*, **94**:745-770.

Murray, H.A. (1938). *Explorations in Personality*. London: Oxford University Press.

Mussen, P.H. (1953). Differences between the TAT responses of negro and white boys. *Journal of Consulting Psychology*, **17**:373-376.

Myers, J.K., and Roberts, B.H. (1959). *Family and Class Dynamics in Mental Illness*. New York: Wiley.

Myrdal, G. (1944). *An American Dilemma*. New York: Harper.

Newsweek Editors (1963). The Negro in America. *Newsweek*, July 29, 1963.

Nobles, W. (1972). African philosophy: Foundations for black psychology. In Jones, R.L. (Ed.), *Black Psychology*. New York: Harper and Row.

Nobles, W. (1973). Psychological research and the black self-concept: A critical review. *Journal of Social Issues*, **29**:11-31.

Nobles, W. (1976). Extended self: Rethinking the so-called negro self-concept. *Journal of Black Psychology*, **2**:15-24.

Nobles, W.W. (1978). Toward an empirical and theoretical framework for defining black families. *Journal of Marriage and the Family*, **40**:668-679.

Nunberg, H. (1931). The synthetic function of the ego. *International Journal of Psychoanalysis*, **12**:123-140.

Osgood, C.E., Suci, G.J., and Tannebaum, P.H. (1957) *The Measurement of Meaning*. Urbana: The University of Illinois Press.

Otnow, D., and Prelinger, E. (1962). An abstract design test of the capacity for intimacy. *Perceptual and Motor Skills*, **15**:645-647.

Page, W.F. (1975). Self-esteem and internal vs external control among black youth in a summer aviation program. *Journal of Psychology*, **89**:307-311.

Parsons, T. (1965). Full citizenship for the American negro: A sociological problem. *Daedalus*, **94**:1009-1054.

Payne, B.F., and Dunn, C.J. (1972). An analysis of the change in self-concept by racial descent. *Journal of Negro Education*, **41**:156-163.

Peters, M.F. (1978). Notes from the guest editor. *Journal of Marriage and the Family*, **40**:655-658.

Pettigrew, T.F. (1964a). Negro American personality: Why isn't it more known? *Journal of Social Issues*, **20**:4-23.

Pettigrew, T.F. (1964b). *A Profile of the American Negro*. New York: Van Nostrand Reinhold.

Pettigrew, T.F. (1964c). Race, emotional illness and intelligence. *Eugenics Quarterly*, **11**:189-215.

Pettigrew, T.F. (1965). Complexity and change in American racial patterns: A sociological view. *Daedalus*, **94**:974-1008.

Phillips, U.B. (1918). *American Negro Slavery: A Survey of the Supply, Employment and Control of Negro Labor as Determined by the Plantation Regime*. New York: Appleton.

Phylon Editors (1964). *Phylon: The Atlantic University Review of Race and Culture*, **25**.

Picou, J.S., Cosby, A.G., Lemke, J.W., and Azuma, H.T. (1974). Occupational choice and perception of attainment blockage: A study of lower class delinquent and non-delinquent black males. *Adolescence*, **9**:289-298.

Pierce, C. (1968). Problems of the negro adolescent in the next decade. In Brody, E. (Ed.) *Minority Group Adolescents in the United States*. Baltimore: Williams and Wilkins.

Piers, G., and Singer, M.B. (1953). *Shame and Guilt*. Springfield, Ill.: Charles C. Thomas.

Polite, C.K., Cochrane, R. and Silverman, B.I. (1974). Ethnic group identification and differentiation. *Journal of Social Psychology*, **92**:149-150.

Polsky, H.W., and Kohn, M. (1959). Participant observation in a delinquent subculture. *American Journal of Orthopsychiatry*, **29**:737-751.

Porter, J.D.R. (1971). *Black Child, White Child: The Development of Racial Attitudes*. Cambridge: Harvard University Press.

Pouissant, A. (1967). Negro self-hate. *New York Times Sunday Magazine*, August 20, 1967.

Pouissant, A., and Atkinson, C. (1972). Black youth and motivation. In Banks, J.A. and Grambs, J.D. (Eds.), *Black Self-Concept*. New York: McGraw-Hill.

Pouissant, A., and Ladner, J. (1968). Black power. *Archives of General Psychiatry*, **18**:385-398.

Powdermaker, H. (1943). The channeling of negro aggression by the cultural process. *American Journal of Sociology*, **48**:750-758.

Powell, G.J. (1973). Self-concept in white and black children. In Willie, C.V., Kramer, B.M., and Brown, B.S. (Eds.), *Racism and Mental Health*. Pittsburgh: University of Pittsburgh Press.

Powell, G.J., and Fuller, M. (1972). The variables for self-concept among Southern black adolescents. *Journal of the National Medical Association*, **64**:522-526.

Prelinger, E. (1958). Identity diffusion and the synthetic function. In Wedge, B.M. (Ed.), *Psychosocial Problems of College Men*. New Haven: Yale University Press.

Prelinger, E. (1960). Discussion of papers by Dr. E. Slocombe and Dr. J. Wilms on problems of ego identity in college students. Unpublished manuscript.

Prelinger, E., and Zimet, C.N. (1964). *An Ego Psychological Approach to Character Assessment.* New York: Macmillan (Glencoe Press).

Prelinger, E., Zimet, C., and Levin, M. (1960). An ego psychosocial scheme for personality assessment. *Psychological Reports,* 7:182f.

Proshansky, H., and Newton, P. (1973). Colour: The nature and meaning of negro self-identity. In Watson, P. (Ed.). *Psychology and Race.* Chicago: Aldine.

Rainwater, L. (1966). Crucible of identity: The negro lower class family. *Daedalus,* 95:172-216.

Rainwater, L., and Yancey, W. (Eds.) (1967). *The Moynihan Report and the Politics of Controversy.* Cambridge: M.I.T. Press.

Rainwater, L. (1970). *Behind Ghetto Walls.* Chicago: Aldine.

Rappaport, D. (1959). A historical survey of psychoanalytic ego psychology. *Psychological Issues,* 1:5-17.

Redding, S. (1951). *On Being Negro in America.* Indianapolis: Bobbs-Merrill.

Reece, C. (1974). Black self-concept. *Child Today,* 3:24026.

Reiss, A.J., Jr. (1961). *Occupations and Social Status.* New York: The Free Press of Glencoe.

Riesman, D. (1950). *The Lonely Crowd.* New Haven: Yale University Press.

Roberts, A., Mosley, K.Y., and Chamberlain, M.W. (1975). Age differences in racial self-identity of young black girls. *Psychological Reports,* 37:1263-1266.

Roe, A. (1956). *The Psychology of Occupations.* New York: Wiley.

Rohrer, J.H., and Edmonson, M.S. (Eds.) (1960). *The Eighth Generation.* New York: Harper.

Rose, A.M. (1956). Psychoneurotic breakdown among soldiers in combat. *Phylon,* 17:61-69.

Rose, A.M. (1965). The negro protest. *Annals of the American Academy of Political and Social Science,* #357.

Rosenberg, M. (1965). *Society and the Adolescent Self Image.* Princeton: Princeton University Press.

Rosenberg, M., and Simmons, R.G. (1971). Functions of children's perceptions of the stratification system. *American Sociology Review,* 36:235-249.

Rosenberg, M., and Simmons, R.G. (1972). *Black and White Esteem: The Urban School Child.* Washington, D.C.:American Sociological Association.

Rosenberg, M. (1979). *Conceiving the Self.* New York: Basic Books.

Ross, A.O. (1962). Ego identity and the social order. *Psychological Monographs,* #542.

Rotter, J.B. (1966). Generalized expectancies for internal versus external control of reinforcement. *Psychological Monographs,* 80 (1, whole No. 609).

Samuels, S.C. (1973). An investigation into the self-concepts of lower and middle class black and white kindergarten children. *Journal of Negro Education,* 42:467-472.

Sanford, N. (Ed.) (1962). *The American College.* New York: Wiley.

Sappington, A., and Grizzard, R. (1975). Self-discrimination responses in black schoolchildren. *Journal of Personality and Social Psychology,* **31**:224-231.

Sarason, I.G., and Ganzer, V.S. (1962). Anxiety reinforcement and experimental instructions in a free verbalization situation. *Journal of Abnormal and Social Psychiatry,* **65**:300-307.

Schachtel, E.G. (1962). On alienated concepts of identity. *Journal of Humanist Psychology,* **1**:110-121.

Schafer, R. (1968). *Aspects of Internalization.* New York: International Universities Press.

Schafer, R., (1973). Concepts of self and identity and the experience of separation, individuation and adolescence. *Psychoanalytic Quarterly,* **42**:12-59.

Schatzman, L., and Strauss, A.L. (1954). Social class and modes of communication. *American Journal of Sociology,* **60**:329-338.

Scher, M. (1967). Negro group dynamics. *Archives of General Psychiatry,* **17**:646-651.

Schilder, P. (1951). *The Image and Appearance of the Human Body.* New York: International Universities Press.

Schofield, J.W. (1978). An exploratory study of the draw-a-person as a measure of racial identity. *Perceptual Motor Skills,* **46**:311-321.

Sclare, A.B. (1953). Cultural determinants in the neurotic negro. *British Journal of Medical Psychology,* **26**:278-288.

Scott, J.F. (1970). Soul brother. *American Journal of Orthopsychiatry,* **40**:216-217.

Secord, P.F. (1959). Stereotyping and favourableness in the perception of negro faces. *Journal of Abnormal and Social Psychiatry,* **59**:309-314.

Shapiro, R. (1963). Adolescence and the psycholgoy of the ego. *Psychiatry,* **26**:77-87.

Shaw, M.E. (1974). The self-image of black and white pupils in an integrated school. *Journal of Personality,* **42**:12-22.

Shilling, F. (1974). Korper Koordination Test fur Kinder (KTK). Weinheim: Beltz Test Gmbh.

Short, J.F., and Strodtbeck, F.L. (1963). Response of gang leaders to status threats. *American Journal of Sociology,* **68**:571-579.

Siegel, S. (1956). The relationship of hostility to authoritarianism. *Journal of Abnormal and Social Psychiatry,* **52**:363-372.

Siegel, S. (1956). *Nonparametric Statistics.* New York: McGraw-Hill.

Silberman, C.E. (1964). *Crisis in Black and White.* New York: Random House.

Sklarew, B.H. (1959). Effects on adolescents of different sexes of early separation from parents. *Psychiatry,* **22**:399-405.

Smith, D., and Mazis, M.B. (1976). Racial self-identification and self-concept by means of unobtrusive measures. *Journal of Social Psychology,* **90**:221-228.

Smith, E.J. (1975). Profile of the black individual in vocational literature. *Journal of Vocational Behavior*, 6:41-59.

Spiegel, L. (1951). A review of the contributions to a psychoanalytic theory of adolescence: Individual aspects. *The Psychoanalytic Study of the Child*, 6:375-393

Spiegel, L. (1958). Comments on the psychoanalytic psychology of adolescence. *The Psychoanalytic Study of the Child*, 13:296-308.

Srole, L., Langner, T.S., Michael, S.T., Opler, M., and Rennie, T. (1962). *Mental Health in the Metropolis: The Midtown Manhattan Study*, Vol. 1 New York: McGraw-Hill.

Stabler, J.R., Johnson, E.E., and Jordan, S.E. (1972). The measurement of children's self-concepts as related to racial membership. *Child Development*, 42:2094-2097.

Staples, R. (1974). The black family revisited: A review and a preview. *Journal of Sociological and Behavioral Sciences*, 19:65-77.

Stein, H., and Cloward, R. (Eds.) (1958). *Social Perspectives on Behavior*. New York: Macmillan (Glencoe Press).

Stephenson, W. (1953). *The Study of Behavior*. Chicago: University of Chicago Press.

Strauss, A.L. (1959). *Mirrors and Masks: The Search for Identity*. New York: Macmillan (Glencoe Press).

Sudarkasa, N. (1975). An exposition on the value premises underlying black family studies. *Journal of the National Medical Association*, 67:235-239.

Sullivan, H.S. (1953). *The Interpersonal Theory of Psychiatry*. New York: Norton.

Symonds, P.N., and Jensen, A.R. (1961). *From Adolescent to Adult*. New York: Columbia University Press.

Taylor, J.A. (1953). A personality scale of manifest anxiety. *Journal of Abnormal and Social Psychiatry*, 8:285-290.

Taylor, R.L. (1976). A psychosocial development among black children and youth: A reexamination. *American Journal of Orthopsychiatry*, 46:4-19.

Taylor, R.L. (1976b). Black youth and psychosocial development—A conceptual framework. *Journal of Black Studies*, 6:353-372.

Taylor, R.L. (1977). The orientation others and value preferences of black college youth. *Social Science Quarterly*, 57:797-810.

Taylor, M.C. & Walsh, E.J. (1979). Explanations of black self-esteem: Some empirical tests. *Social Psychiatry Quarterly*, 42:242-253.

Teahan, J.E., and Podany, E.C. (1974). Some effects of films of successful blacks on racial self-concept. *International Journal of Social Psychiatry*, 20:274-280.

Teplin, L.A. (1976). A comparison of racial/ethnic preferences among Anglo, Black, and Latino children. *American Journal of Orthopsychiatry*, 46:702-709.

Terry, R.L., and Evans, J.E. (1972). Class versus race discrimination attributed to self and others. *Journal of Psychology*, 80:183-187.

Thomas, A., and Sillen, S. (1972). *Racism and Psychiatry*. New York: Bruner/Mazel.

Trowbridge, N.T. (1969). Project IMPACT research report 1968-1969. U.S. Office of Educational Cooperative Research Report.

Trowbridge, N.T. (1970). The measurement. In Rowson, J.P. (Ed.) IMPACT 70. Des Moines, Iowa, Polk County Education Services, 55-74.

Trowbridge, N. (1972). Self-concept and socio-economic status in elementary school children. *American Education Research Journal*, 9:525-538.

Trowbridge, N., Trowbridge, L., and Trowbridge, L. (1972). Self-concept and socio-economic status. *Child Study Journal*, 2:123-143.

United States Employment Service. (1971). Manual for the U.S. Employment Service non-reading aptitude battery, Form A. Washington, D.C.: United States Department of Labor.

Valentine, C.A. (1971). Deficit, difference, and bicultural models of Afro-American behavior. *Harvard Educational Review*, Reprint Series, No. 5, 1-21.

Ward, S.H., and Braun, J. (1972). Self-esteem and racial preference in black children. *American Journal of Orthopsychiatry*, 42:644-647.

Warheit, G.J., Holzer, C.E. III, Avey, S.A. (1975). Race and mental illness: An epidemiologic update. *Journal of Health and Social Behavior*, 16:243-256.

Wax, D.E. (1972). Self-concept in negro and white preadolescent delinquent boys. *Child Study Journal*, 2:175-184.

Weaver, E.K. (1956). Racial sensitivity among negro children. *Phylon*, 17:52-60.

Wedge, B.M. (1958). *Psychosocial Problems of College Men*. New Haven: Yale University Press.

Wender, P.H. (1968). Vicious and virtuous circles: The role of deviation amplifying feedback in the origin and precipitation of behavior. *Psychiatry*, 31:309-324.

Wheelis, A. (1958). *The Quest for Identity*. New York: Norton.

White, J. (1972). Toward a black psychology. In Jones, R.L. (Ed.) *Black Psychology*, New York: Harper and Row.

White, R.K. (1947). Black boy: A value analysis. *Journal of Abnormal and Social Psychology*, 42:440-461.

White, R.W. (1952). *Lives in Progress*. New York: Dryden Press.

White, R.W. (1963). Sense of interpersonal competence: Two case studies and some reflections on origins. In White, R.W., *The Study of Lives*. Englewood Cliffs, New Jersey: Prentice-Hall.

White, R.W. (1963). *The Study of Lives*. New York: Atherton.

Whyte, W.F. (1955). *Street Corner Society*. Chicago: University of Chicago Press.

Whyte, W.F. (1958). A slum sex code. In Stein, H., and Cloward, R. (Eds.), *Social Perspectives on Behavior* New York: Macmillan (Glencoe Press).

Willie, C.V. (1974). The black family and social class. *American Journal of Orthopsychiatry*, 44:50-60.

Witmer, H. (1956). *New Perspectives in Delinquency*. Washington, D.C.: U.S. Government Printing Office.

Wilstad, A.M. (1961). Identity conflicts in disturbed adolescent girls. *Smith College Studies in Social Work*, **32**:20-37.

Wohlwill, J.F. (1973). *The study of behavioral development*. New York: Academic Press.

Wolf, E., Geds, J.E., and Terman, D.M. (1972). On the Adolescent Processes and Transformation of the Self. *Journal of Youth and Adolescence*, **1**:257-272.

Woodmanese, J.J., and Cook, S.W. (1967). Dimensions of verbal racial attitudes: Their identification and measurement. *Journal of Personality and Social Psychiatry*, **7**:240-250.

Wright, R. (1945). *Black Boy*. New York: Harper.

Wylie, R.C. (1961). *The Self-Concept: A Critical Survey of Pertinent Research Literature*. Lincoln: University of Nebraska Press.

Yancey, W.L., Rigsby, L., and McCarthy, J.D. (1972). Social position and self-evaluation: The relative importance of race. *American Journal of Sociology*, **78**:338-359.

Yando, R., Seitz, V., and Zigler, E., (1979). *Intellectual and Personality Characteristics of Children*. Hillsdale, N.J.: Erlbaum.

Young, F.M. (1959). Response of juvenile delinquents to the TAT. *Journal of Genetic Psychology*, **88**:251-259.

Zirkel, P., and Gable, R.K. (1977). The reliability and validity of various measures of self-concept among ethnically different adolescents. *Measurement and Evaluation in Guidance*, **10**:48-54.

Zirkel, P.A., and Moses, E.G. (1971). Self-concept and ethnic group membership among public school students. *American Education Research Journal*, **8**:253-265.

Name Index

Subject Index